AMERICAN MATHEMATICAL SOCIETY
TRANSLATIONS

Series 2

Volume 142

Ten Papers Translated from the Russian

by

M. B. Abalovich
D. S. Dzhumabaev
V. Z. Feĭnberg
V. Fleischer
S. S. Goncharov
B. N. Khabibullin
M. G. Kreĭn
N. Ya. Krupnik
B. Ya. Levin
L. V. Nosov
E. G. Nosova
A. A. Nudel′man
E. B. Rabinovich
G. I. Russu
V. Yu. Sazonov
D. I. Sviridenko

AMERICAN MATHEMATICAL SOCIETY
PROVIDENCE, RHODE ISLAND

Translation edited by BEN SILVER

1980 *Mathematics Subject Classification* (1985 *Revision*). Primary 03B60, 16A40, 26C99, 30D20, 41A10, 41A50, 46E30, 46H10, 46H15, 46J20, 47D30, 47H15, 47H17, 49D15, 52A45, 68N05, 68U05; Secondary 03C70, 03F05, 46B15, 47A53, 68N15, 70J10.

Library of Congress Cataloging-in-Publication Data
Ten papers translated from the Russian/by M. B. Abalovich...[et al.; translation edited by Ben Silver].
p. cm. – (American Mathematical Society translations, ISSN 0065-9290; ser. 2, v. 142)
Includes bibliographies.
ISBN 0-8218-3122-4
1. Mathematics–1961- I. Abalovich, M. B. II. Silver, Ben. III. Title: 10 papers translated from the Russian. IV. Series.
QA3.A572 vol. 142
[QA7]
510 s–dc19 89-371
[510] CIP

The paper used in this book is acid-free and falls within the guidelines
established to ensure permanence and durability. ♾

Contents

Russian Contents*

В. Ю. Сазонов, Принцип коллекции и квантор существования, Вычисл. системы (Новосибирск), 1985, № 107, 30–39

Л. В. Носов, Е. Г. Носова, Е. Б. Рабинович и В. З. Фейнберг, Оптимальные алгоритмы решения задач взаиморасположения и покрытия плоских многоугольных фигур, Весці АН БССР. Сер. фіз.-мат. н., 1985, № 4, 28–36

В. Фляйшер, Вполне плоские моноиды, Уч. зап. Тарт. ун-та, 1982, № 610, 38–52

М. Г. Крейн, Б. Я. Левин и А. А. Нудельман, О специальном представлении многочлена, полохительного на системе замкнутых интервалов, Физ.-техн. ин-т низк. температур АН УССР. Препр., 1984, № 28

Б. Н. Хабибуллин, Разложение целых функций конечного порядка на эквивалентные множители, Вопр. аппроксимации функций веществ. и комплекс. перемин. Уфа, 1983, 161–181

Г. И. Руссу, О совпадении идеалов строго сингулярных операторов и ϕ-допустимых возмущений в пространствах Лоренца, Исслед. по функц. анализу и дифференц. уравнениям. Мат. н. Кишинев, 1984, 112–121

М. Б. Абалович и Н. Я. Крупник, Топология на множестве максимальных идеалов банаховых PI-алгебр, Исследования по функц. анализу и дифференц. уравнениям. Мат. н. Кишинев, 1984, 3–11

*The American Mathematical Society scheme for transliteration of Cyrillic may be found at the end of index issues of *Mathematical Reviews*.

Amer. Math. Soc. Transl.
(2) Vol. **142**, 1989

The Collection Principle and the Existential Quantifier

UDC 510.6

V. YU. SAZONOV

Abstract. A nonclassical sequent calculus RBC ("relativization to recursive bounds with collection") is presented, one of whose axioms is the well-known set-theoretic collection principle for arbitrary formulas A: $\forall x < a \exists y A \to \exists b \forall x < a \exists y < b A$. By means of proof theory for RBC the following analogue of Harrop's theorem is obtained: from any inference $\mathrm{RBC} \vdash \exists y < A(x, y)$ it is possible to construct effectively the term $t(x)$ and inference $\mathrm{RBC} \vdash \exists y < t(x) A(x, y)$. It is established that an extension using the Σ-collection principle of classical theories whose axioms and theorems contain only bounded quantifiers is conservative.

Bibliography: 5 titles.

Introduction

In this paper we shall present a certain nonclassical sequent calculus RBC ("relativization to recursive bounds with collection"), one of the axioms of which is the well-known set-theoretic collection principle [1] for arbitrary formulas A:

$$\forall x < a \exists y A \to \exists b \forall x < a \exists y < b A. \qquad \text{(Coll)}$$

By means of the theory of proofs for RBC we have obtained an analogue of Harrop's theorem [5] or the E-property for intuitionistic logic. (See Corollary 1 of the relativization theorem; this result was mentioned earlier in [2] and [4]). In fact, for every derivation $\mathrm{RBC} \vdash \exists y A(x, y)$, one can effectively construct a term $t(x)$ and a derivation $\mathrm{RBC} \vdash \exists y < t(x) A(x, y)$.

By the same method as one proves the violation of the law of the excluded middle in formal arithmetic with the E-property (that is, with reference to the nonderivability of consistency), one can show that there is no (completely) classical and arbitrary logical calculus with the property mentioned above. Another proof for this is presented in the following section. However, if we

---

1980 *Mathematics Subject Classification* (1985 *Revision*). Primary 03B60; Secondary 03F05, 03C70.

Translation of Vychisl. Sistemy Vyp. 107(1985), 30–39; MR **87e**:03116.

do not remain entirely in the language of formulas with unbounded quantifiers, then we essentially obtain the classical logical calculus RBΔ_0 (without Coll!). It is shown that the collection principle for Σ-formulas A is interpretable in RBΔ_0 by means of a suitable relativization of unbounded existential quantifiers. This implies the conservativeness of extensions of classical theories by means of a Σ-collection principle, the axioms of which contain only bounded quantifiers (unbounded outer universal quantifiers are, as usual, omitted; see Corollary 2 of the relativization theorem). In this connection, one uses a natural generalization of a theorem on cut elimination [2].

The deficiency of the calculus RBC is the absence in its language of implication and the unbounded universal quantifier. But their introduction, with preservation of the results mentioned above, requires a significant revision of the approach presented below for the construction of the calculus (and its semantics). This will be done in another paper.

Our investigation is essentially connected with the search for a formal apparatus in support of a program that dispenses with potential realizability and thus relativizes constructions and arguments within recursive bounds, that is, respects these bounds and does not deviate from them. In [2]–[4], there is a more detailed discussion of this program and the connection of it and the collection principle with the logical analysis of concepts and problems of the theory of complexity. In actuality, the collection principle directly expresses the requirement of relativization of quantifiers with respect to recursive bounds, and is a logical principle rather than a special axiom characterizing the domain of individuals. The indicated analogue of the E-property of the calculus RBC is obviously also in accord with this aim and even, to a significant degree, determines it.

Interpreting $<$ as the membership relation $\in$, we can consider the calculus RBC (or a variant of it) as a logic for a certain set theory, in which (in contrast with ZF or KP [1]) all arguments and constructions are "relativized to sets". The method of avoiding the paradoxes that is adopted in the traditional axiomatic set theories seems rather artificial. On the one hand, instead of unrestricted comprehension $\{x|A(x)\}$, one is permitted to use only the restricted form, $\{x \in a|A(x)\}$. On the other hand, both in the comprehension scheme (in the case of ZF) and in the foundation scheme (in the case of KP), A can contain unbounded quantifiers. More consistent and thorough-going is the restriction of these schema with respect to Δ_0-comprehension (as in KP) or with respect to Δ_0-foundation. The (Δ_0- or Σ-) collection principle is justified by the discussion above (or the indicated result about its conservativeness). Instead of unrestricted foundation we now would have to assume a rule for introducing set-theoretic recursive functions. We particularly mention the axiom of the existence of power-sets (and arithmetic powers) as unacceptable, even if the finiteness of all sets is postulated, in contrast with, say, the axiom of the existence of the transitive closure of an arbitrary set. On the other hand, nothing stops us from "locally" strengthening such a set theory by an axiom guaranteeing the existence of one transitive set—a model for ZF. However, a more detailed investigation of these questions requires separate consideration.

The author is grateful to D. I. Sviridenko for many useful observations.

§1. The calculus RB

We fix a primitive predicate $x < y$, signifying that x lies within the bound y. In the general case, we make no assumptions about the transitivity or reflexivity of $<$. Depending on the interpretation, this can be a linear order or even the membership relation. The form of the axioms imposed on $<$ makes $\in$ the preferred reading of $<$. We also introduce a function $[x, y]$, an analogue of an unordered pair $(x < [x, y], y < [x, y])$. Then $[x] \rightleftharpoons [x, x]$ plays the role of a one-element set, and the relation $x \subseteq y \rightleftharpoons \forall v < x\ (v < y)$ plays the role of set-theoretic inclusion. Finally, we fix an analogue of the union operation $\bigsqcup_{x<y} t(x, y, \bar{z})$, binding the variable x in the term t:

$$v < y \to t(v, y, \bar{z}) \subseteq \bigsqcup_{x<y} t(x, y, \bar{z}).$$

By repeated application these operations can be extended to the case of lists of variables $\bar{x}$ and $\bar{y}$. We set $\bigsqcup x \rightleftharpoons \bigsqcup_{v<x} v, x \sqcup y \rightleftharpoons \bigsqcup [x, y]$, and $[t|\bar{x} < \bar{y}] = \bigsqcup_{\bar{x}<\bar{y}} [t]$.

Also using other primitive predicate and function symbols, we can form, in the usual way, first-order formulas. Formulas of the language RB contain, by definition, only the connectives $\wedge, \vee, \neg, \exists$, and bounded universal quantifiers of the form $\forall x < t \cdots$, or, more precisely, $\forall x(\neg x < t \vee \cdots)$. Similarly, bounded existential quantifiers are regarded as quantifiers of the form $\exists x < t \cdots$, that is, $\exists x(x < t \wedge \cdots)$. Naturally, in both cases it is assumed that x does not occur free in the term t. Formulas containing only restricted quantifiers are called *Δ_0-formulas*. An occurrence in a formula of an unbounded quantifier $\exists$ is said to be *accessible* if it is not within the scope of the connective $\neg$. Formulas in which all occurrences of (unbounded) quantifiers $\exists$ are accessible (respectively, inaccessible) form the class Σ (respectively, N). A *sequent* is an expression $\Gamma \to \Delta$, where Γ and Δ are arbitrary finite lists of formulas, possibly empty. If $\Gamma, \Delta \subseteq \Delta_0$ (Σ), then $\Gamma \to \Delta$ is called a Δ_0 (Σ) sequent.

Axioms of RB:

$$\to x < [x, y];$$

$$\to y < [x, y];$$

$$v < y \to t(v, y, \bar{z}) \subseteq \bigsqcup_{x<y} t(x, y, \bar{z});$$

$A \to A$, where A is any formula of the language.

The *rules of inference* of RB are the rule of substitution of terms for free variables in a sequent

$$\frac{\Gamma(\bar{x}) \to \Delta(\bar{x})}{\Gamma(\bar{s}) \to \Delta(\bar{s})}$$

as well as all the rules of Gentzen's sequent calculus LK with cuts (see [5]), applied only to sequents of the language RB, with a restriction on the rule

$$\frac{A, \Gamma \to \Delta}{\Gamma \to \Delta, \neg A} \qquad (\to \neg)$$

which is taken to be applicable if $A \in N$ or if the list Δ is empty.

In particular, the rules for $\forall$ have the form

$$\Gamma \to \Delta, \neg x < t \vee A / \Gamma \to \Delta, \forall x < tA$$

and

$$\neg s < t \vee A(s), \Gamma \to \Delta / \forall x < tA(x), \Gamma \to \Delta.$$

If we admit the rule $(\to \neg)$ only for Δ_0-formulas A or only for formulas A in N, then we obtain, respectively, the calculi RB_0 and RB_1. We note that the fragment $\mathrm{RB}\Delta_0$ of RB which involves only Δ_0-formulas is essentially the usual classical logic in a language with bounded quantifiers plus axioms for $<, [,], $ and $\bigsqcup$. One defines $\mathrm{RB}\Sigma$ in similar fashion. Obviously, $\mathrm{RB}\Delta_0 \subseteq \mathrm{RB}\Sigma \subseteq \mathrm{RB}_0 \subseteq \mathrm{RB}_1 \subseteq \mathrm{RB}$, where the inclusions are understood in the sense of the strength (of special cases) of the axioms and rules of inference.

The following sequents are easily derived in $\mathrm{RB}\Delta_0$:

$$\begin{aligned}
&\to x < [x];\\
&v < x \to v \subseteq \bigsqcup x;\\
&\to x \subseteq x \sqcup y \wedge y \subseteq x \sqcup y;\\
&\bar{v} < \bar{y} \to t(\bar{v}, \bar{y}, \bar{z}) \subseteq \bigsqcup_{\bar{x} < \bar{y}} t(\bar{x}, \bar{y}, \bar{z});\\
&\bar{v} < \bar{y} \to t(\bar{v}, \bar{y}, \bar{z}) < [t(\bar{x}, \bar{y}, \bar{z}) | \bar{x} < \bar{y}].
\end{aligned}$$

By the *principle of $\mathscr{A}$-collections* we mean the scheme

$$\forall x < a \exists y A \to \exists b \forall x < a \exists y < bA. \qquad (\mathscr{G}\text{-Coll})$$

where $A \in \mathscr{A}$ and $\mathscr{A}$ is a fixed class of formulas. The complete collection principle Coll corresponds to the class $\mathscr{A}$ of all formulas. We set $\mathrm{RBC} \rightleftharpoons \mathrm{RB} + \mathrm{Coll}$. Below we shall show that

$$\mathrm{RB}_0 + N\text{-Coll} \vdash \mathrm{Coll} \quad \text{and } \mathrm{RB}\Sigma + \Delta_0\text{-Coll} \vdash \Sigma\text{-Coll}. \qquad (*)$$

In actuality, in these cases we shall even derive formally more general collection principles, in which lists of variables $\bar{x}$, $\bar{a}$, $\bar{y}$ are involved instead of the variables x, a, y.

The complete collection principle, combined with classical logic, enables us to prove the realizability (more precisely, the everywhere definability) of very rapidly growing functions, in comparison with functions given by terms of the language. For this it suffices that the graph $F(x, y)$ of such a function satisfies the following natural condition for all a:

$$\exists b \forall x < a\,(\exists y F(x, y) \supset \exists y < b F(x, y)) \supset \forall x < a\,\exists y F(x, y).$$

In fact, on the basis of the law of the excluded middle, the collection principle, the indicated property of F, and the axiom $x < [x]$, we successively obtain (for $a = [x]$):

$$\begin{aligned}
&\forall x < a\,\exists y(\exists z F(x, y) \supset F(x, y)),\\
&\exists b \forall x < a\,\exists y < b(\exists z F(x, z) \supset F(x, y)),\\
&\forall x < a\,\exists y F(x, y),\\
&\forall x \exists y F(x, y).
\end{aligned}$$

For example, one proves in this way the realizability of the powers 2^n (Theorem 7.3.1 in [2]) and even the "increasing towers"

$$\left.2^{2^{\cdot^{\cdot^{\cdot^{2}}}}}\right\} n \text{ times}$$

in the theory $T_0 + \text{Coll}$, where T_0 consists of all true quantifier-free formulas composed out of polynomially computable predicates and functions.

In order to exclude such consequences of the collection principle (see the Introduction also), we shall, from now on, consider the calculus RB instead of classical logic.

We denote by A^t the result of replacing in A all accessible occurrences of (unbounded) quantifiers $\exists x$ by $\exists x < t$. As in the Kripke–Platek theory [1], it is not difficult to show by induction on the structure of A that

$$\begin{aligned}
&\mathrm{RB}_0 \vdash A^v \to A;\\
&\mathrm{RB}_0 \vdash \exists v A^v \to A;\\
&\mathrm{RB}_0 \vdash u \subseteq v \wedge A^u \to A^v \quad \text{(monotonicity)};\\
&\mathrm{RB}_0 + N\text{-Coll} \vdash A \to \exists v A^v.
\end{aligned}$$

In addition, for Σ-formulas A, instead of RB_0 one can take $\mathrm{RB}\,\Sigma$, and instead of N-Coll the scheme Δ_0-Coll. For example, the latter assertion, for A of the form $\forall x < t A_0$, is obtained by the following steps:

$$\begin{aligned}
\forall x < t A_0 &\to \forall x < t\,\exists u A_0^u \to \exists w \forall x < t\,\exists u < w A_0^u\\
&\to \exists v \forall x < t\,\exists u \subseteq v A_0^u \to \exists v \forall x < t A_0^v.
\end{aligned}$$

Here, the first implication is based upon the inductive hypothesis, the second on the collection principle, the third on the change from w to $v = \bigsqcup w$, and the fourth on the monotonicity of A^u with respect to u.

In actuality, the scheme $A \to \exists v A^v$ is equivalent to the collection principle. In fact, $\forall \bar{x} < \bar{a}\,\exists \bar{y} B$ implies $\exists v \forall \bar{x} < \bar{a}\,\exists \bar{y} < v B^v$, whence, by monotonicity,

$\exists v \forall \bar{x} < \bar{a} \exists \bar{y} < vB$. In this way, we have arrived at the proposition $(*)$ indicated above.

§2. The basic result

The sequent $S^{t(v)} \rightleftharpoons \Gamma^v \to \Delta^{t(v)}$ is said to be the *relativization* of the sequent $S \rightleftharpoons \Gamma \to \Delta$ to the term t with respect to the variable v (v does not occur in S).

THEOREM (on relativization). a) *For any derivation* $\mathrm{RBC}+\{S_i\} \vdash S$ *and terms* $\{t_i(v)\}$, *one can effectively construct a term* $t(v)$ *and a derivation* $\mathrm{RBC}+\{S_i^{t_i(v)}\} \vdash S^{t(v)}$.

b) *In addition, if the first derivation is carried out in* $\mathrm{RB}_1+\mathrm{Coll}+\{S_i\}$ *or in* $\mathrm{RB}\Sigma+\Sigma\text{-}Coll+\{S_i\}$ (*and* $S_i, S \in \Sigma$), *then the second can be carried out in* $\mathrm{RB}_1+\{S_i^{t_i(v)}\}$ *or, respectively, in* $\mathrm{RB}\Delta_0+\{S_i^{t_i(v)}\}$ ($S_i^{t_i(v)} \in \Delta_0$).

COROLLARY 1. *If* $\mathrm{RBC} \vdash\ \to \exists x A$, *then, for some term* t, $\mathrm{RBC} \vdash \exists x < tA^t$ *and* $\mathrm{RBC} \vdash \exists x < tA$.

Part b) of the theorem, as well as Corollary 2, which follows from it, can be considered as a kind of relative justification for the collection principle.

COROLLARY 2. *If* $\mathrm{LK}+\{S_i\}+\Sigma\text{-}Coll \vdash S$ *and* S_i, $S \in \Sigma$, *then, for any* $\{t_i\}$, *there exist* t *and an* LK *derivation* S^t *from the axioms for* $[\,,\,]$ *and* $\bigsqcup$ *and* $\{S^{t_i}\}$, *consisting only of* Δ_0*-formulas; that is, a derivation in* $\mathrm{RB}\Delta_0+\{S_i^{t_i}\}$.

The proof of Corollary 2 is easily obtained with the help of the following theorem for $Ax = \Sigma$, since it guarantees, in that case, the existence of an LK derivation consisting of Σ-formulas, that is, a derivation of S in $\mathrm{RB}\Sigma+\{S_i\}+\Sigma\text{-Coll}$.

GENERALIZED THEOREM ON CUT ELIMINATION [2]. *Let* Ax *be a set of sequents of the calculus* LK, *closed with respect to substitution of terms. Then cuts with respect to formulas not occurring in the sequents of* Ax *can be eliminated from any derivation in* $\mathrm{LK}+Ax$.

PROOF OF THE RELATIVIZATION THEOREM. Obviously, without loss of generality, we can assume that S is obtained from $\{S_i\}$ by an application of exactly one rule of inference of RB, or that S is an axiom of RBC. We shall examine the most typical cases.

1. $S \rightleftharpoons$ Coll. Setting $t \rightleftharpoons v \sqcup [v]$, we obtain a derivation of Coll^t:

$$\frac{\forall x < a\, \exists y < vA^v \to v < v \sqcup [v] \wedge \forall x < a\, \exists y < vA^{v \sqcup [v]}}{\forall x < a\, \exists y < vA^v \to \exists b < v \sqcup [v] \wedge \forall x < a\, \exists y < bA^{v \sqcup [v]}}.$$

2. Substitution: $\Gamma(\bar{x}) \to \Delta(\bar{x}) / \Gamma(\bar{s}) \to \Delta(\bar{s})$. For any relativization $\Gamma^v(\bar{x}) \to \Delta^{t(v,\bar{x})}(\bar{x})$, we have the required derivation of the relativization of the sequent $\Gamma(\bar{s}) \to \Delta(\bar{s})$ to $t(v, \bar{s})$:

$$\frac{\Gamma^v(\bar{x}) \to \Delta^{t(v,\bar{x})}(\bar{x})}{\Gamma^v(\bar{s}) \to \Delta^{t(v,\bar{s})}(\bar{s})}.$$

3. A cut

$$\frac{\Gamma \to \Delta, A \qquad A, \Pi \to \Lambda}{\Gamma, \Pi \to \Delta, \Lambda}$$

can be converted into the following derivation (assuming, for simplicity, that $\mathrm{RB}\Delta_0 \vdash \to v \subseteq t_1(v)$):

$$\dfrac{\to v \sqsubseteq t_1 \qquad \dfrac{\Gamma^v \to \Delta^{t_1}, A^{t_1} \qquad \dfrac{A^v, \Pi^v \to \Lambda^{t_2(v)}}{A^{t_1}, \Pi^{t_1} \to \Lambda^{t_2(t_1)}}}{\Gamma^v, \Pi^{t_1} \to \Delta^{t_1}, \Lambda^{t_2(t_1)}}}{\dfrac{\Gamma^v, \Pi^v \to \Delta^{t_1}, \Lambda^{t_2(t_1)}}{\Gamma^v, \Pi^v \to \Delta^{t_1 \sqcup t_2(t_1)}, \Lambda^{t_1 \sqcup t_2(t_1)}}}$$

4. The rule $(\to \wedge)$

$$\frac{\Gamma \to \Delta, A \qquad \Gamma \to \Delta, B}{\Gamma \to \Delta, A \wedge B}$$

can be converted into the derivation

$$\dfrac{\dfrac{\Gamma^v \to \Delta^{t_1}, A^{t_1}}{\Gamma^v \to \Delta^{t_1 \sqcup t_2}, A^{t_1 \sqcup t_2}} \qquad \dfrac{\Gamma^v \to \Delta^{t_2}, B^{t_2}}{\Gamma^v \to \Delta^{t_1 \sqcup t_2} B^{t_1 \sqcup t_2}}}{\Gamma^v \to \Delta^{t_1 \sqcup t_2}, \ (A \wedge B)^{t_1 \sqcup t_2}.}$$

5. The rule $(\to \exists$ unbounded)

$$\frac{\Gamma \to \Delta, A(s)}{\Gamma \to \Delta, \exists x A(x)}$$

can be transformed into the derivation

$$\dfrac{\dfrac{\Gamma^v \to \Delta^t, A^t(s)}{\Gamma^v \to \Delta^{t \sqcup [s]}, s < t \sqcup [s] \wedge A^{t \sqcup [s]}(s)}}{\Gamma^v \to \Delta^{t \sqcup [s]}, \exists x < t \sqcup [s] \wedge A^{t \sqcup [s]}(x)}$$

6. The rule $(\exists$ unbounded $\to)$

$$\frac{A(x), \Gamma \to \Delta}{\exists x A(x), \Gamma \to \Delta}$$

can be transformed into the derivation

$$\dfrac{\dfrac{A^v(x), \Gamma^v \to \Delta^{t(v,x)}}{x < v \wedge A^v(x), \Gamma^v \to \Delta^{T(v)}}}{\exists x < v \wedge A^v(x), \Gamma^v \to \Delta^{T(v)},}$$

where $T(v) \rightleftharpoons \bigsqcup_{x<v} t(v, x)$.

7. The rule $(\exists$ bounded $\to)$

$$\frac{x < s \wedge A(x), \Gamma \to \Delta}{\exists x < s \wedge A(x), \Gamma \to \Delta}$$

can be transformed into the derivation

$$\dfrac{\dfrac{x < s \wedge A^v(x), \Gamma^v \to \Delta^{t(v,x)}}{x < s \wedge A^v(x), \Gamma^v \to \Delta^{T_1(v)}}}{\exists x < s \wedge A^v(x), \Gamma^v \to \Delta^{T_1(v)},}$$

where $T_1(v) = \bigsqcup_{x<s} t(v,x)$.

8. The rule ($\to \forall$ bounded)

$$\frac{\Gamma \to \Delta, \neg x < s \vee A(x)}{\Gamma \to \Delta, \forall x < sA(x)}$$

can be transformed into the derivation

$$\frac{\dfrac{\Gamma^v \to \Delta^{t(v,x)}, \neg x < s \vee A^{t(v,x)}(x)}{\Gamma^v \to \Delta^{T_1(v)}, \neg x < s \vee A^{T_1(v)}(x)}}{\Gamma^v \to \Delta^{T_1(v)}, \forall x < sA^{T_1(v)}(x)}$$

where $T_1(v) \rightleftharpoons \bigsqcup_{x<s} t(v,x)$.

9. The rule ($\neg \to$)

$$\frac{\Gamma \to \Delta, A}{\neg A, \Gamma \to \Delta}$$

can be transformed into the derivation

$$\frac{\dfrac{\Gamma^v \to \Delta^t, A^t \quad A^t \to A}{\Gamma^v \to \Delta^t, A}}{\neg A, \Gamma^v \to \Delta^t}.$$

10. The rule ($\to \neg$)

$$\frac{A, \Gamma \to \Delta}{\Gamma \to \Delta, \neg A}$$

can be transformed as follows: in the case where A does not contain accessible occurrences of unbounded quantifiers $\exists$, the derivation becomes

$$\frac{A, \Gamma^v \to \Delta^t}{\Gamma^v \to \Delta^t, \neg A},$$

and, in the case where Δ is empty, the derivation becomes

$$\frac{\dfrac{A \to \exists u A^u \qquad \dfrac{\dfrac{A^v, \Gamma^v \to}{A^u, \Gamma^v \to}}{\exists u A^u, \Gamma^v \to}}{A, \Gamma^v \to}}{\Gamma^v \to \neg A.}$$

We note that the last subcase is the only place in the proof where one uses the collection principle and where non-Δ_0-formulas could appear.

BIBLIOGRAPHY

1. Jon Barwise, *Admissible sets and structures*, Springer-Verlag, 1975.

2. V. Yu. Sazonov, *A logical approach to the problem "$P = NP$?"*, Mathematical Foundations of Computer Science, 1980 (Proc. Ninth Sympos., Rydzyna, 1980), Lecture Notes in Computer Sci., vol. 88, Springer-Verlag, 1980, pp. 562–575.

3. ——, *Polynomial computability and recursivity in finite domains*, Elektron. Informationsverarb. Kybernet. **16** (1980), 319–323.

4. ——, *The checking problem from the point of view of logic*, Logic and the Foundations of Mathematics (Abstracts Eighth All-Union Conf. Logic and Methodology of Science, Pelanga, 1982; B. M. Kedrov et al., editors), Vil′nyus. Gos. Univ., Vilnius, 1982, pp. 77–80. (Russian)

5. Gaisi Takeuti, *Proof theory*, North-Holland, 1975.

Translated by E. MENDELSON

Amer. Math. Soc. Transl.
(2) Vol. **142**, 1989

Optimal Algorithms for Solving Problems of Arrangement and Covering of Plane Polygonal Figures

UDC 519.8; 681.3; 621.3.049.77

L. V. NOSOV, E. G. NOSOVA, E. B. RABINOVICH
AND V. Z. FEĬNBERG

In this paper we consider problems that arise naturally in the automation of projection and the manufacture of integrated circuits. These problems include the covering of plane polygonal figures by elementary figures (rectangles, trapezoids, triangles), the computation of the areas of plane polygonal figures, and the determination of the arrangement of plane contours. The needs of microelectronics make the development of effective algorithms for solving these problems urgent, since most modern integrated circuits may contain tens of thousands of elements, which makes it impossible in practice to use algorithms even with quadratic fulfillment time. Known algorithms for solving these problems include both very simple sorting algorithms [1], [2], whose fulfillment time is necessarily no better than quadratic, and algorithms in which we use methods of data processing such as sorting, fusion, and so on [3]–[6]. However, in the worst case the latter have quadratic fulfillment time.

Our aim is to construct algorithms for solving the problems mentioned above in a time comparable with the sorting time, and to prove that these algorithms are optimal. The construction of the algorithms is carried out in the spirit of the methods of so-called computational geometry [8], which is understood as the development of effective solutions of geometric problems connected with computations. Let us proceed to a statement of these problems.

1980 *Mathematics Subject Classification* (1985 *Revision*). Primary 68U05, 52A45.
Translation of Vestsī Akad. Navuk BSSR Ser. Fiz.-Mat. Navuk **1985**, no. 4, 28–36; MR**87a**: 05050.

On the plane we are given a set of simple (that is, without intersecting nonadjacent sides) polygons,

$$A_1, A_2, \ldots, A_k, \tag{1}$$

whose boundaries (contours) are pairwise disjoint, and so for any A_i and A_j, $i \neq j$, of the set (1) either $A_i \subset A_j$, or $A_j \subset A_i$, or $A_i \cap A_j = \varnothing$.

With the set (1) we associate a graph T whose vertices are the polygons (1), and a pair (A_i, A_j), $i \neq j$, is an edge of T if and only if $A_i \supset A_j$, and from the inclusion $A_i \supseteq A_s \supseteq A_j$ it follows that either $s = i$ or $s = j$. Obviously, T is a forest consisting of rooted trees, since each polygon is the root of a subtree whose vertices are polygons contained in it.

As usual, we define the level of a vertex with respect to a forest consisting of rooted trees: the level of a vertex with respect to a forest is defined as its level with respect to the tree containing it; the root of a tree has level 1; the remaining vertices have level one more than that of the subtree with this root that contains them.

Thus, the level of a polygon in T is one more than the number of polygons containing it. Figure 1a shows a set of polygons, and Figure 1b the forest T corresponding to it.

On the plane we define a set A as follows: a point a belongs to A if and only if a belongs to an odd number of polygons of the set (1). Consider some point $a \in A$ and the smallest (with respect to inclusion) polygon containing a. From the definition of the set A this polygon obviously has an odd level in T. Clearly, boundary points of the polygons A, and only they, are boundary points of A. Thus, A is a plane polygonal figure whose internal points are situated between the boundaries of polygons of odd and even level. To be definite we shall assume that A is closed. Figure 1c shows a polygonal figure A corresponding to the polygons in Figure 1a.

The starting point for all the problems listed below is a set of polygons (1); that is, neither the forest T nor the polygonal figure A is given explicitly.

PROBLEM 1. To construct a covering of a polygonal figure A by trapezoids and triangles in such a way that no two figures of this covering have interior points in common.

The important ones in practice (see [3] and [6]) are plane polygonal figures with sides parallel to the coordinate axes. We shall call such figures rectangles.

PROBLEM 2. To construct a covering of a rectangular figure A by rectangles in such a way that no two rectangles have interior points in common.

Problems 1 and 2 are central in systems of automation of programming for display generators, which form a constituent part of modern systems of automated projection and the manufacture of integrated circuits. These problems have been considered in a number of papers ([1], [3]–[6]).

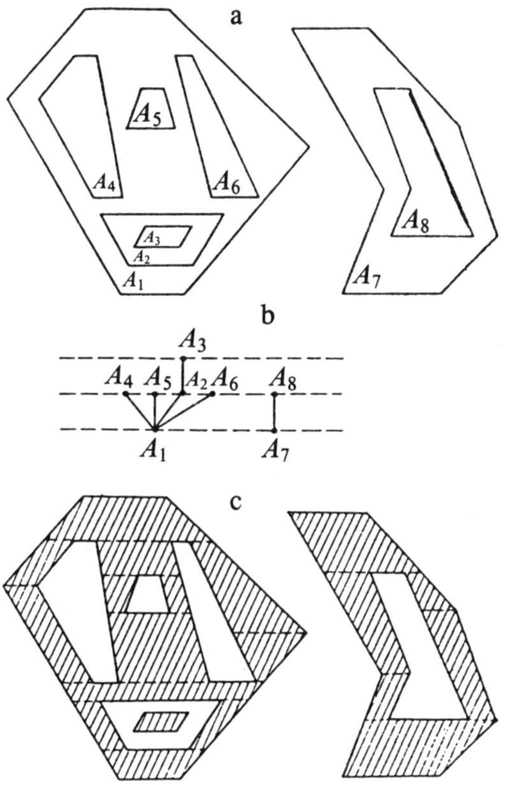

FIGURE 1. Definition of polygonal figure A and a covering of it by trapezoids: a) the set of polygons; b) the forest T; c) the polygonal figure A and a covering of it by trapezoids.

PROBLEM 3. To construct the forest T (to determine all its edges). This was posed in [2] as one of the fundamental problems in systems of automation of programming for display generators.

PROBLEM 4. To compute the area of a plane polygonal figure A. This problem is a natural generalization of the problem of computing the area of a polygon. If (x_i, y_i), $i = 0, 1, \dots, n-1$, are the coordinates of the vertices of a polygon, then its area is computed trivially in time $O(n)$. If for each polygon of the set (1) we knew the parity of its level in T, then Problem 4 would be solved in a time that is linear in the total number of vertices of the polygons. In fact, we would compute the sum of the areas of the polygons of odd level and subtract from the resulting number the sum of the areas of the polygons of even level.

In this paper we give a uniform scheme of computation, which is the foundation for algorithms that solve Problems 1–4 in time $O(N \log N)$, where N is the total number of vertices of the polygons (1), and we prove that $O(N \log N)$ is the lower bound of the time needed to solve these problems. We have therefore proved the following result.

THEOREM. *Problems* 1–4 *are solved in time* $O(N \log N)$, *and* $O(N \log N)$ *is the lower bound of the time needed to solve them.*

Let us proceed to describe the scheme of computation that is fundamental for solving Problems 1–4. On the plane we consider a set R of pairwise disjoint nonhorizontal intervals. We can introduce an order relation on this set in a natural way. We shall say that two intervals $r_1, r_2 \in R$ are *comparable* if there is a horizontal line l intersecting both of them. We have $r_1 \leq r_2$ if the point of intersection of r_1 with l lies to the left of the point of intersection of r_2 with l. It is clear that in this way a partial order relation is introduced on the set R. It extends to the set of nonhorizontal sides of the polygons (1) if by definition we suppose that the intervals r_1 and r_2 in Figure 2a are not comparable and $r_1 < r_2$ in Figure 2b. Without loss of generality we assume that among the nonhorizontal sides of the polygons (1) there are no pairs of intervals r_1, r_2 as shown in Figure 2b. From what follows it will be clear that the inclusion of additional operations in the algorithms makes it possible to process such pairs of intervals without asymptotic growth of fulfillment time.

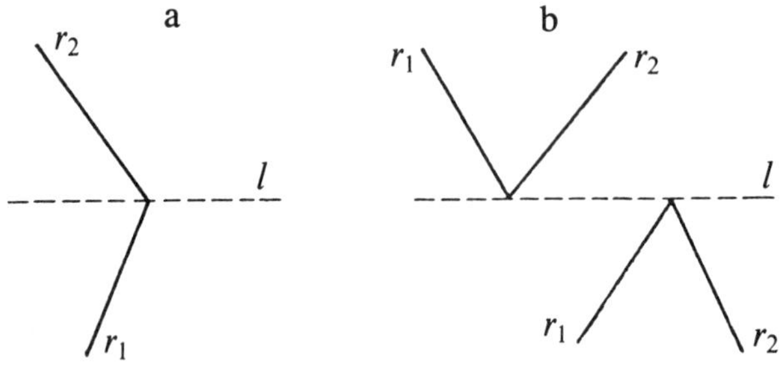

FIGURE 2. The definition of an order relation on a set of intervals.

The main step in the computation scheme, which we give below, is the successive renewal of the linearly ordered subset $R(l)$ of the set R. Suppose that the horizontal line l is displaced in the plane from the lowest point of intersection with sides of the polygons (1) to the uppermost point. The sides whose lower ends lie on l are included in $R(l)$, while the sides whose upper ends lie on L are excluded from $R(l)$. All the inclusions and exclusions will be carried out from left to right: if in the motion along l from left to right we meet a side of a polygon whose lower end lies on l, we include it in $R(l)$, if we meet a side whose upper end lies on l, we exclude it from $R(l)$, and if we meet a pair of sides as in Figure 2a we first exclude r_1 from $R(l)$ and then include r_2. Suppose that in a given position of l we have gone along it from the extreme left to the extreme right of its points of intersection with sides of the polygons (1). Then the set $R(l)$ consists of the nonhorizontal sides of the polygons (1) that intersect l, except for those sides whose upper ends lie on l. We denote this state of $R(l)$ by $\overline{R}(l)$. We can show that the interval of l whose left end is a point of intersection of l with an odd interval r_1 of

$\overline{R}(l)$ and whose right end is a point of intersection of l with the interval r_2 ($r_1 < r_2$) adjacent to r in $\overline{R}(l)$ belongs to A. Let l_1 be the position of l in the upward motion such that r_1 and r_2 become consecutive elements of $\overline{R}(l)$ (r_1 odd, r_2 even, $r_1 < r_2$). We move l upwards from the position l_1 until r_1 and r_2 cease to be successive elements of $\overline{R}(l)$ (that is, until either the interval r, $r_1 < r < r_2$, is not included in $R(l)$ or one of the intervals is not excluded from $R(l)$). We denote this last position of l by l_2. Then the figure obtained as a result of the intersection of four half-planes—to the right of the line on which r_1 lies, to the left of the line on which r_2 lies, above l_1, and below l_2—is a trapezoid with horizontal bases and belongs entirely to A. We denote this trapezoid by (r_1, r_2, l_1, l_2). The set of such figures is a covering of A satisfying the conditions of Problem 1.

We now turn to the choice of a suitable data structure to represent the ordered set $R(l)$. To this end, use was made in [3]–[7] of a linear list. The renewal of $R(l)$ is carried out successively by means of the operations of exclusions from the linear list (of intervals whose upper ends lie on l) and fusion (with the set of intervals whose lower ends lie on l). It is not difficult to see that this leads to the fact that the algorithms in [3]–[7] are satisfied at worst in time $O(N^2)$.

In [9] in the construction of an algorithm that makes it possible to determine in time $O(N \log N)$ whether in a given set of N intervals in the plane there is at least one pair of intersecting intervals, a binary balanced tree [10] was used as the data structure. The application of this data structure for the representation of the ordered set $R(l)$ enables us to carry out in time at worst $O(\log N)$ the operations of inclusion and exclusion of an interval r from $R(l)$ and the determination of the intervals of $R(l)$ closest to r on the left and the right. We denote these operations respectively by INSERT$(r, R(l))$, REMOVE$(r, R(l))$, LEFT$(r, R(l))$, and RIGHT$(r, R(l))$.

We note that we can establish whether an interval r is an even or odd element of $\overline{R}(l)$ by determining its parity in $R(l)$ for the inclusion of r in $R(l)$, since at the instant of including r in $R(l)$ the ordered sets $R(l)$ and $\overline{R}(l)$ coincide up to the element r. For the determination of the parity of an element under inclusion of it in a binary balanced tree, see [11].

Let us turn to the formal presentation of the algorithm for solving Problem 1. We assume that the polygons (1) are given in a rectangular Cartesian coordinate system and that $x_1(r)$, $y_1(r)$ are the coordinates of the lower end of r, and $x_2(r)$, $y_2(r)$ are the coordinates of the upper end.

ALGORITHM 1. We sort the ends of the nonhorizontal sides of the polygons (1) lexicographically with respect to y and to x. Let $\{P(I)\}$, $I = 1, \dots, 2N$, be the set obtained as a result of the sorting. We shall assume that the upper end of r_1 in Figure 2a precedes the lower end of r_2 in this ordered set.

We put $I := 1$ and $R(l) := \varnothing$.

1. Let r be the interval whose end is $P(I)$. We check which end (upper or lower) of r is $P(I)$.

If $P(I)$ is the lower end of r, we carry out the operation INSERT$(r, R(l))$ and find $r_1 :=$ LEFT$(r, R(l))$.

If r_1 is an even element of $\overline{R}(l)$ or $y_1(r) = y_1(r_1)$, we proceed to carry out 2. Otherwise we obtain a trapezoid (r_1, r_2, l_1, l_2) from the covering of the polygonal figure A, where $r_2 :=$ RIGHT$(r, R(l))$, and l_1 and l_2 are the lines $y = y_1(r_1)$ and $y = y_1(r)$ respectively. We replace $x_1(r_1)$, $y_1(r_1)$ and $x_1(r_2)$, $y_1(r_2)$ by the coordinates of the points of intersection of r_1 and r_2 respectively with the line $y = y_1(r)$ and proceed to carry out 2.

If $P(I)$ is the upper end of r, we check whether r degenerates to a point, that is, $y_1(r) = y_2(r)$.

If this is so, we carry out the operation REMOVE$(r, R(l))$ and proceed to carry out 2. Otherwise we obtain a trapezoid (r_1, r_2, l_1, l_2) from the covering of the polygonal figure, where l_1 and l_2 are the lines $y = y_1(r)$ and $y = y_2(r)$ respectively, and r_1 and r_2 are defined as follows.

If r is an even element of $\overline{R}(l)$, then $r_2 := r$ and $r_1 :=$ LEFT$(r, R(l))$. In this case we replace $x_1(r_1)$ and $y_1(r_1)$ by the coordinates of the point of intersection of r_1 with the line $y = y_2(r)$.

If r is an odd element of $\overline{R}(l)$, then $r_1 := r$ and $r_2 :=$ RIGHT$(r, R(l))$. In this case we replace $x_1(r_2)$ and $y_1(r_2)$ by the coordinates of the point of intersection of r_2 with the line $y = y_2(r)$.

After this we carry out the operation REMOVE$(r, R(l))$ and proceed to carry out 2.

2. We put $I := I + 1$. If $I > 2N$, then the algorithm completes the work. If not, we proceed to carry out 1.

PROPOSITION 1. *Algorithm* 1 *solves Problem* 1 *in time* $O(N \log N)$.

PROOF. A preliminary sorting of the ends of the sides of the polygons (1) requires time $O(N \log N)$ [11]. For each interval at the general stage 1 we carry out either the operation INSERT$(r, R(l))$ or the operation REMOVE$(r, R(l))$ and also either the operation LEFT$(r, R(l))$ or the operation RIGHT$(r, R(l))$ or both operations. Since each operation requires no more than $O(\log N)$ of time, and the computation of the coordinates of the vertices of the trapezoids of the covering A requires a time bounded by some constant, the total time for carrying out Algorithm 1 does not exceed $O(N \log N)$.

In Algorithm 1 the determination of the trapezoids is carried out only when r is included in $R(l)$ or excluded from $R(l)$, so we have the following result.

PROPOSITION 2. *Algorithm* 1 *constructs a covering of the polygonal figure* A *consisting of* $O(N)$ *figures.*

PROPOSITION 3. *Problem* 2 *is solved in time* $O(N \log N)$.

PROOF. In fact, if Algorithm 1 has in the input a set of rectangular polygons which determines a rectangular figure A, then it constructs a covering of it by rectangles. This covering satisfies the condition of Problem 2.

PROPOSITION 4. *The area of figure A is computed in time* $O(N \log N)$.

PROOF. We construct a covering of A by means of Algorithm 1. By Proposition 2 the number of figures in this covering is $O(N)$. Since the time needed to compute the area of each figure of the covering is bounded by some constant, the computation of the sum of the areas of all these figures requires time $O(N)$. Consequently, the total time spent in computing the area of A does not exceed $O(N \log N)$.

We consider an algorithm for solving Problem 3 below. Meanwhile we prove that $O(N \log N)$ is a lower bound of the time needed to solve Problems 1–4. For this we need to consider the following auxiliary problem.

PROBLEM 5. To determine whether there is at least one pair of intersecting polygons in the set (1).

From the definition of the forest T it follows that this problem is equivalent to the following: whether there is at least one polygon in the set (1) whose level with respect to T is different from 1.

PROPOSITION 5. *Problem* 5 *is solved in time* $O(N \log N)$, *and* $O(N \log N)$ *is the lower bound of the time needed to solve it.*

PROOF. It is easy to show that each polygon of the set (1) has level 1 with respect to T if and only if the area of the polygonal figure A is equal to the sum of the areas of the polygons (1). We can compute the sum of the areas of the polygons in time $O(N)$, and by Proposition 4 we can compute the area of A in time $O(N \log N)$.

We consider on the number axis the set of natural numbers $\{x_i\}$, $i = 1, \dots, N$. For each point x_i we construct the square B_i with side $2/3$ and center x_i. Obviously, $B_i \cap B_j = \varnothing$ if $x_i \neq x_j$, and $B_i \supset B_j$ if $x_i = x_j$ and $i < j$. Then this set of squares falls under the definition of the set (1), and if we solve Problem 5 for it, we thereby determine whether among the numbers x_i, $i = 1, \dots, N$, at least one pair are equal. This problem was considered in [12], where it was proved that $O(N \log N)$ is the lower bound of the time needed to solve it. In solving it we have spent time $O(N)$ on the construction of the set of squares plus the time needed to solve Problem 5. Hence it follows that the time needed to solve Problem 5 cannot be less than $O(N \log N)$.

COROLLARY. $O(N \log N)$ *is the lower bound of the time needed to solve Problems* 1–4.

The proof of Problem 4 follows immediately from the proof of Proposition 5, and the proof of Problem 3 follows from the fact that any algorithm for solving it also obviously solves Problem 5.

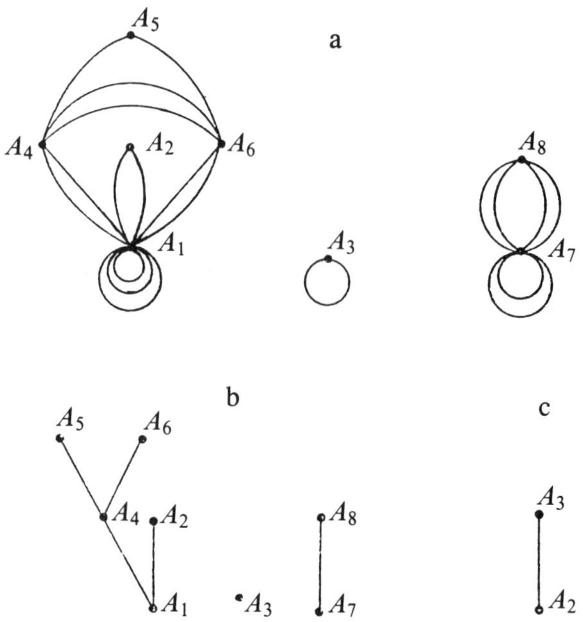

FIGURE 3. The definition of the edge specification of the forest T: a) the multigraph M; b) the forest M_1; c) the forest M_2.

We consider the algorithm that solves Problem 1 or 2, and construct a covering of A consisting of N' figures. Since this algorithm computes the coordinates of each figure of the covering, the time for carrying it out cannot be less than $O(N')$. Therefore if the time for carrying out the algorithm is less than $O(N \log N)$, then the number of figures of the covering that it constructs is less than $O(N \log N)$. But then, by summing the areas of the figures of this covering, we can compute the area of A in less time than $O(N \log N)$. This contradicts the fact that $O(N \log N)$ is the lower bound of the time needed to solve Problem 4. Consequently, there is no algorithm that solves Problem 1 or 2 in less time than $O(N \log N)$.

To prove the theorem it now remains to show that there is an algorithm for constructing the forest T in time $O(N \log N)$.

We note that each of the two lateral sides of the figures of the covering that Algorithm 1 constructs is the whole or part of a side of some polygon of the set (1). This makes it possible to define a multigraph with loops M whose vertices are the polygons (1), and each figure of this covering determines an edge (A_i, A_j), where A_i and A_j are polygons to whose contours its lateral sides belong. Then the vertices of a connected component of M are vertices of some shrub of the forest T that grows from a vertex of odd level.

By Proposition 2 the sum of the number of vertices and the number of edges of the multigraph M is equal to $O(N)$. Since the connected components of a multigraph can be constructed in a time that is linear in the sum of the number of vertices and the number of edges, the set of all shrubs of T that grow from vertices of odd level can be constructed in time $O(N)$.

To determine the shrubs of T that grow from vertices of even level it is sufficient to supplement the set of polygons (1) by a rectangle A_0 in which all the polygons (1) are contained, and to repeat all the constructions carried out above.

Taking account of the fact that the time for carrying out Algorithm 1 is equal to $O(N \log N)$, and all the additional constructions require time $O(N)$, we see that Problem 3 is solved in time $O(N \log N)$. This completes the proof of the theorem.

To construct an algorithm for solving Problem 3 we shall define not a multigraph M but its spanning forest M_1 (the family of spanning trees of the connected components of M). Each tree of M_1 is a rooted tree, and its root is the root of the corresponding shrub of T. The algorithm given below finds in parallel the arcs of the forest M_1 whose trees correspond to the shrubs of T that grow from vertices of odd level, and the arcs of the forest M_2 whose trees correspond to shrubs of T that grow from vertices of even level (see Figure 3).

ALGORITHM 2. In each polygon of the set (1) we mark the extreme left of the lower nonhorizontal sides. As in Algorithm 1, we sort the set of ends of the nonhorizontal sides of the polygons.

We put $I := 1$ and $R(l) = \varnothing$.

1. Let r be the interval whose end is $P(1)$. We check which end of r (upper or lower) is $P(1)$.

If $P(1)$ is the upper end of r, we carry out the operation REMOVE($r, R(l)$) and proceed to carry out 2.

If $P(1)$ is the lower end of r, we carry out the operation INSERT($r, R(l)$). If r is not marked, we proceed to carry out 2. Otherwise we put i equal to the number of the polygon whose side is r.

If r is the extreme left of $R(l)$, we define a fictitious edge (A_i, A_0) of the forest M_2 (A_0 is a fictitious rectangle in which all the polygons (1) are contained) and proceed to carry out 2. Otherwise:

a) we find $r_1 :=$ LEFT($r, R(l)$);

b) we put j equal to the number of the polygon with side r_1; and

c) we define the edge (A_i, A_j), which is an edge of M_1 if r is an even element of $R(l)$ or an edge of M_2 otherwise.

After this we proceed to carry out 2.

2. We put $I := I + 1$. If $I > 2N$, we carry out 3. Otherwise we carry out 1.

3. We determine the connected components of M_1 and M_2 and find the roots of the resulting trees, regarding the edge (A_i, A_j) as directed from A_i to A_j. We exclude from consideration the fictitious connected component of M_2 whose root is A_0. The remaining components of M_2 and the connected components of M_1 determine all the shrubs of the forest T.

REMARK. A theorem on the optimality of the algorithms we have considered can be proved on the assumption that $O(k) = O(N)$, that is, the number of polygons of the set (1) is comparable with the total number of their vertices (this is the most important case for the solution of practical problems of automation of projection and the manufacture of integrated circuits).

BIBLIOGRAPHY

1. G. G. Kazentsov, L. B. Osipov, and V. M. Shchemelinin, Èlektronnaya Promyshlennost′ **1974**, no. 6, 84–87. (Russian)

2. A. L. Stempkovskiĭ, Mikroèlektronika **1976**, no. 9, 365–372. (Russian)

3. G. È. Shiro, L. B. Osipov, and V. I. Volkov, Èlektronnaya Tekhnika Ser. 11, **1976**, vyp. 3, 71–76. (Russian)

4. L. B. Osipov and G. È. Shiro, Èlektronnaya Tekhnika Ser. 10, **1978**, vyp. 2(8), 89–103. (Russian)

5. A. L. Stempkovskiĭ, Èlektronnaya Tekhnika Ser. 3, **1978**, vyp. 6, 74–80. (Russian)

6. E. G. Nosova, A. G. Sverdlov, and V. Z. Feĭnberg, *Algorithms for partitioning plane figures in systems for computer design of integrated circuits*, Vestsī Akad. Navuk BSSR Ser. Fiz.-Mat. Navuk **1978**, no. 5, 16–23. (Russian)

7. L. V. Nosov, E. B. Rabinovich, and V. Z. Feĭnberg, *An algorithm for the arrangement of plane geometric figures*, Vychisl. Tekhn. i Mashinostr. (Minsk) **1979**, no. 2, 13–18. (Russian) R. Zh. Mat. **1980**, 3B 1251.

8. G. T. Toussaint, *Pattern recognition and geometrical complexity*, Proc. Fifth Internat. Conf. Pattern Recognition (Miami Beach, Fla., 1980), IEEE, New York, 1980, pp. 1324–1347.

9. Michael Ian Shamos and Dan Hoey, *Closest-point problems*, Sixteenth Annual Sympos. Foundations of Computer Sci. (Berkeley, Calif., 1975), IEEE Computer Soc., Long Beach, Calif., 1975, pp. 151–162.

10. G. M. Adel′son-Vel′skiĭ and E. M. Landis, *An algorithm for the organization of information*, Dokl. Akad. Nauk SSSR **146** (1982), 263–266; English transl. in Soviet Math. Dokl. **3** (1962).

11. Donald E. Knuth, *The art of computer programming*. Vol. 3: *Sorting and searching*, Addison-Wesley, 1973.

12. David P. Dobkin and Richard J. Lipton, Tech. Rep. No. 42, Dept. Computer Sci., Yale Univ., New Haven, Conn., 1975, pp. 191–208.

Translated by E. J. F. PRIMROSE

Amer. Math. Soc. Transl.
(2) Vol. **142**, 1989

Completely Flat Monoids

V. FLYAĬSHER [FLEISCHER]

In this article we consider flat polygons over monoids. Using the concept of the tensor product of polygons, Kil′p [2] and Stenström [8] introduced two different concepts of flatness for a polygon. Kil′p defined a left polygon B to be *flat* if, as in the theory of modules over rings, the functor of tensor multiplication by B preserves monomorphisms (i.e., imbeddings). Stenström defined a polygon B to be flat if the functor of tensor multiplication by B preserves equalizers and co-universal squares. As shown in [3], a polygon which is flat in the Stenström sense is flat in the Kil′p sense (the converse is false in general). Therefore, it is natural to call flatness in the Stenström sense strong flatness [3].

A monoid over which all left and all right polygons are flat is said to be *completely flat*. A result on describing completely flat monoids which are sheafs of groups was announced in [4]. The main result in the present article is a complete description of arbitrary completely flat monoids. Here the consideration of one more type of flatness for a polygon, which we call *weak flatness*, turns out to be useful. Weak flatness of a polygon is also a natural analogue of flatness of a module over a ring. It is known ([5], §5.4) that a left module B over a ring R is flat if and only if the functor $\otimes B$ preserves imbeddings of right ideals of R, regarded as R-modules, into R. We say that an arbitrary left polygon B over a monoid S is *weakly flat* if the functor of tensor multiplication by B preserves imbeddings of the form $T \to S$, where T is an arbitrary right ideal of S, regarded as an S-polygon. It is clear that every flat polygon is weakly flat. The converse is false in general (the author has an appropriate example). Along with a description of completely flat monoids we obtain a description of monoids over which all left polygons are weakly flat, as well as monoids over which all left cyclic Rees polygons are flat. We remark that a description of the latter was known only in the commutative

1980 *Mathematics Subject Classification* (1985 *Revision*). Primary 16A50.
Translation of Tartu Riikl. Ül. Toimetised No. 610 (1982), 38–52; MR **85k**: 20178.

case ([3], Theorem 2). The main result of this article implies that all left (and right) polygons over an inverse monoid are flat. The description of monoids over which all left polygons are flat which was given in the class of monoids with central idempotents (see [3], Theorem 1) thereby extends to the class of monoids with commuting idempotents. We note that in the general case the question of describing monoids over which all left polygons are flat remains open.

Everywhere below, S is a monoid with identity denoted by 1. A set A is called a *right polygon* over S, or a *right S-polygon*, if the product $as \in A$ is defined for any $a \in A$ and $s \in S$, with $(as)t = a(st)$ and $a \cdot 1 = a$ for all $a \in A$ and $s, t \in S$. *Left S-polygons* are defined similarly. In what follows, S-polygons will simply be called *polygons*. Let A be a right and B a left polygon. The definition of the tensor product of A and B is analogous to that of the tensor product of modules over a ring. Namely, in the direct product $A \times B$ of the sets A and B we take the smallest equivalence relation Θ for which

$$(as, b)\ \Theta\ (a, sb)$$

for all $a \in A$, $b \in B$, and $s \in S$. The quotient set $(A \times B)/\Theta$ is called the *tensor product* of the polygons A and B, and is denoted by $A \otimes B$. It is easy to see that $(a_1, b_1)\ \Theta\ (a_2, b_2)$ if and only if there is a finite chain of pairs in $A \times B$ such that the first pair coincides with (a_1, b_1), the last pair concides with (a_2, b_2), and the transition from each pair of the chain to the next takes place by transposition of an element in S, i.e., from a pair (as, b) to the pair (a, sb), or conversely. (The transition from one pair to another by transposition of an element in S will be denoted by the symbol $\sim$.) The equivalence class to which the pair (a, b) belongs is denoted by $a \otimes b$. If we fix a left polygon B and assign to each right polygon A the set $A \otimes B$, then we get a functor $\otimes B$ from the category of right polygons to the category of sets. A left polygon B is said to be *flat* if every monomorphism of arbitrary right polygons $\varphi\colon A \to C$ induces an imbedding $\varphi \otimes I_B\colon A \otimes B \to C \otimes B$ of sets. In other words, a left polygon B is said to be *flat* if for any monomorphism $\varphi\colon A \to C$ of arbitrary right polygons A and C and any a, $a' \in A$ and b, $b' \in B$ the equality $\varphi(a) \otimes b = \varphi(a') \otimes b'$ in the tensor product $C \otimes B$ implies the equality $a \otimes b = a' \otimes b'$ in $A \otimes B$.

The following assertion gives a criterion for flatness of a polygon.

THEOREM 1. *A left polygon B is flat if and only if for any cyclic right polygon $A = aS$ and any subpolygon A' of it the imbedding $A' \to A$ induces an imbedding $A' \otimes B \to A \otimes B$.*

PROOF. The necessity is obvious.

Sufficiency. Let $\iota\colon C \to D$ be an imbedding of right polygons C and D. We show that this imbedding induces an imbedding $C \otimes B \to D \otimes B$. The polygons C and $\iota(C)$ will be identified, i.e., C will be assumed to be a subpolygon of D. Let $\mathfrak{A}$ be the collection of all polygons M such that $C \subseteq M \subseteq D$ and such that the imbedding $C \to M$ induces an imbedding $C \otimes B \to M \otimes B$. Clearly $A \neq \varnothing$, since $C \in \mathfrak{A}$. We show that $\mathfrak{A}$ is closed with respect to taking the union of an increasing chain. Let M_1, $M_2, \dots$ be an increasing chain of polygons in $\mathfrak{A}$, i.e.,

$$M_1 \leq M_2 \leq \cdots \leq M_n \leq \cdots. \tag{1}$$

Then $M = M_1 \cup M_2 \cup \cdots$ is a subpolygon of D, and $C \leq M \leq D$. To prove that $M \in \mathfrak{A}$ it suffices to prove that if the equality

$$m \otimes b = m' \otimes b' \tag{2}$$

holds in $M \otimes B$ for some m, $m' \in C$ and b, $b' \in B$, then this equality also holds in $C \otimes B$. Suppose that (2) holds in $M \otimes B$. This means that there exists a finite chain L of pairs of elements in $M \times B$ joining the pairs (m, b) and (m', b'). Let $m = m_1, m_2, \dots, m_k = m'$ be the elements of M which occur in the first positions of the pairs in this chain. They are finite in number, since L is finite. Then there is a polygon M_n in the increasing chain (1) which contains all these elements. This means that L can be regarded as being in the set $M_n \times B$, i.e., $m \otimes b = m' \otimes b'$ in $M_n \otimes B$. Since $M_n \in \mathfrak{A}$, this implies that $m \otimes b = m' \otimes b'$ in $C \otimes B$. Thus, $M \in \mathfrak{A}$.

By Zorn's lemma, this gives us that there is a maximal polygon $P \in \mathfrak{A}$. If $P = D$, then $D \in \mathfrak{A}$ and it is thereby proved that the imbedding $C \to D$ induces an imbedding $C \otimes B \to D \otimes B$. We assume that $P \neq D$ and show that this leads to a contradiction. Let $a \in D \times P$ and consider the polygon $P \cup aS$. We show that $P \cup aS \in \mathfrak{A}$. Assume that (2) holds in $(P \cup aS) \otimes B$ for some m, $m' \in C$ and b, $b' \in B$. This means that there exists a finite chain L of pairs

$$(m, b) = (m_1, b_1),\ (m_2, b_2), \dots, (m_k, b_k) = (m', b') \tag{3}$$

in $(P \cup aS) \times B$ joining the pairs (m, b) and (m', b'). If in all the pairs in L the elements m_i $(1 \leq i \leq k)$ belong to P, then (2) holds in $P \otimes B$, and hence also in $C \otimes B$. Assume now that among the elements $m_1, \dots, m_k$ in the chain (3) there are elements in aS which do not belong to P. Let m_j $(1 < j < k)$ be such an element, i.e., $m_j \in aS$ and $m_j \notin P$. We show that there exists a chain L' of pairs in $(P \cup aS) \times B$ joining (m, b) and (m', b') and not containing (m_j, b_j), and that all the pairs contained in L' and not in L belong to $P \times B$.

Let i be the minimal index such that $i \leq j$, m_i, $m_{i+1}, \dots, m_j \in aS$, and $m_{i-1} \notin aS$. If there is no such i, then let $i = 1$. On the other hand, let r be a maximal index such that $r \geq j$, m_j, $m_{j+1}, \dots, m_r \in aS$, and $m_{r+1} \notin aS$. If there is no such r, then let $r = k$. We show first that m_i, $m_r \in aS \cap P$. This will show that both $i \neq j$ and $r \neq j$, since $m_j \notin P$.

If $i = 1$, then $m_i = m_1 = m \in P$, i.e., $m_i \in P \cap aS$. If $i \neq 1$, then $m_{i-1} \notin aS$. Assume that the pair (m_i, b_i) is obtained from (m_{i-1}, b_{i-1}) by transposition of an element $x \in S$ from left to right, i.e.,

$$(m_{i-1}, b_{i-1}) = (m_i x, b_{i-1}) \sim (m_i, x b_{i-1}) = (m_i, b_i).$$

Then $m_{i-1} = m_i\dot{x} \in aS$, which contradicts the choice of the index i. Hence, (m_i, b_i) is obtained from (m_{i-1}, b_{i-1}) by transposition of some element $y \in S$ from right to left, i.e.,

$$(m_{i-1}, b_{i-1}) = (m_{i-1}, yb_i) \sim (m_{i-1}y, b_i) = (m_i, b_i).$$

Consequently, $m_i = m_{i-1}y \in P$, and thus $m_i \in P \cap aS$.

The assertion $m_r \in P \cap aS$ can be proved similarly. To do this it suffices to rewrite the chain (3) in reverse order and use the above arguments for the element m_r. Thus, the chain L has a subchain K

$$(m_i, b_i),\ (m_{i+1}, b_{i+1}), \dots, (m_j, b_j), \dots, (m_r, b_r)$$

of pairs in $aS \times B$ with m_i, $m_r \in P \cap aS$. Consequently, $m_i \otimes b_i = m_r \otimes b_r$ in $aS \otimes B$. By assumption, the latter equality must hold also in $(P \cap aS) \otimes B$, since $P \cap aS$ is a subpolygon of a cyclic polygon. Hence, there exists a chain K',

$$(m_i, b_i) = (m'_1, b'_1),\ (m'_2, b'_2), \dots, (m'_l, b'_l) = (m_r, b_r),$$

of pairs in $(P \cap aS) \times B$ joining (m_i, b_i) and (m_r, b_r). We now consider the chain L' obtained by replacing the subchain K by K' in L,

$$\begin{aligned}(m, b) = (m_1, b_1), \dots, (m_i, b_i) = (m'_1, b'_1), \dots,\\ (m'_l, b'_l) = (m_r, b_r), \dots, (m_k, b_k) = (m', b').\end{aligned}$$

This chain does not contain the pair (m_j, b_j), and all its pairs not in L are in K', and hence in $P \times B$. Since the number of pairs (m_j, b_j) in L with $m_j \notin P$ is finite, by using the procedure finitely many times we get a chain L^* of pairs in $P \times B$ joining (m, b) and (m', b'), i.e., $m \otimes b = m' \otimes b'$ in $P \otimes B$. The last equality holds also in $C \otimes B$, because $P \in \mathfrak{A}$. We have thus shown that $P \cup aS \in \mathfrak{A}$. In view of $a \notin P$, this contradicts the maximality of P, and the theorem is proved.

Thus, Theorem 1 shows that a left polygon B is flat if and only if the functor $\otimes B$ preserves imbeddings into an arbitrary cyclic right polygon, while for a weakly flat polygon U the functor $\otimes U$, by definition, preserves imbeddings into a free cyclic right polygon. We remark that this situation has already been encountered in the study of injective polygons (see [1] and [7]), where, along with the concept of an injective polygon as an analogue of the concept of an injective module, the concept of a weakly injective polygon arises: a polygon satisfying the analogue of the Baer criterion.

LEMMA 2. *Let B be an arbitrary left polygon. The equality $s \otimes b = s' \otimes b'$ holds in the tensor product $S \otimes B$ for some s, $s' \in S$ and b, $b' \in B$ if and only if $sb = s'b'$.*

PROOF. *Sufficiency.* Let $sb = s'b'$. Then there exists a chain

$$(s, b) = (1 \cdot s, b) \sim (1, sb) = (1, s'b') \sim (s', b'),$$

which implies that $s \otimes b = s' \otimes b'$.

Necessity. Let $s \otimes b = s' \otimes b'$. This means that there exists a finite chain of pairs in $S \times B$, namely,

$$(s, b) = (s_1, b_1), (s_2, b_2), \dots, (s_k, b_k) = (s', b'), \tag{4}$$

each pair of which, beginning with the second, is obtained from the preceding pair by transposition of an element in S. Suppose that the pair (s_2, b_2) is obtained from (s_1, b_1) by transposition of an element $x \in S$ from left to right, i.e.,

$$(s_1, b_1) = (s_2 x, b_1) \sim (s_2, x b_1) = (s_2, b_2).$$

Then $s_1 b_1 = (s_2 x) b_1 = s_2 (x b_1) = s_2 b_2$. A similar result obtains if the transposition is from right to left. Thus, $sb = s_1 b_1 = s_2 b_2$. Similarly, $s_2 b_2 = s_3 b_3$, and so on. Since the chain (4) is finite, we get that $sb = s_1 b_1 = s_2 b_2 = \cdots = s_k b_k = s' b'$, and the lemma is proved.

For any element s in a regular monoid let s^* denote an element in this monoid such that $ss^*s = s$. The next lemma bears an auxiliary nature.

LEMMA 3. *Let S be a regular monoid, and B a left polygon such that for any $b \in B$ the equality $xb = yb$ for some $x, y \in S$ implies that there is a $z \in xS \cap yS$ for which $zb = xb = yb$. Let $A = aS$ be an arbitrary cyclic right S-polygon. If $as \otimes b = at \otimes g$ in $A \otimes B$ for some $s, t \in S$ and $b, g \in B$ and the corresponding chain in $A \times B$ contains two transpositions, the first from left to right, then the equality $as \otimes b = at \otimes g$ holds also in $(asS \cup atS) \otimes B$.*

PROOF. Suppose that $as \otimes b = at \otimes g$ in $A \otimes B$ and that the corresponding chain of pairs in $A \times B$ has the form

$$(as, b) = (as_1 x_1, b) \sim (as_1, x_1 b) = (as_1, x_2 g) \sim (as_1 x_2, g) = (at, g).$$

This implies that $x_1 b = x_2 g$. Denote this element by g', i.e., $b' = x_1 b = x_2 g$. Since S is regular, there are elements $x_1^*, x_2^* \in S$ such that $x_1 x_1^* x_1 = x_1$ and $x_2 x_2^* x_2 = x_2$. Then

$$x_1 x_1^* b' = x_1 x_1^* x_1 b = x_1 b = x_2 g = x_2 x_2^* x_2 g = x_2 x_2^* b'.$$

By assumption, this implies that there exists an element $z = x_1 x_1^* u = x_2 x_2^* v$ such that $zb' = x_1 x_1^* b' = x_2 x_2^* b'$. Then there exists a chain of pairs in $(asS \cup atS) \times B$:

$$\begin{aligned}(as, b) &= (as_1 x_1 x_1^* x_1, b) \sim (as_1 x_1 x_1^*, x_1 b) = (as_1 x_1 x_1^*, x_1 x_1^* b') \\ &= (as_1 x_1 x_1^*, x_1 x_1^* u b') \sim (as_1 x_1 x_1^* x_1 x_1^* u, b') \\ &= (as_1 x_1 x_1^* u, b') = (as_1 x_2 x_2^* v, b') = (as_1 x_2 x_2^* x_2 x_2^* v, b') \\ &\sim (as_1 x_2 x_2^*, x_2 x_2^* v b') = (as_1 x_2 x_2^*, x_2 x_2^* b') \\ &= (as_1 x_2 x_2^*, x_2 x_2^* x_2 g) \sim (as_1 x_2, g) = (at, g).\end{aligned}$$

Thus, $as \otimes b = at \otimes g$ in $(asS \cup atS) \otimes B$, and the lemma is proved.

For arbitrary elements $x, y \in S$ we denote by $\lambda_{x,y}$ and $\rho_{x,y}$ minimal left and right congruences of S, respectively, which put x and y together in the same class.

THEOREM 4. *For an arbitrary monoid S the following are equivalent:*

1) *all left S-polygons are weakly flat;*

2) *all finitely generated left S-polygons are weakly flat;*

3) *all cyclic left S-polygons are weakly flat;*

4) *S is a regular monoid, and for any x, $y \in S$ there is an element $z \in xS \cap yS$ such that $(x, z) \in \lambda_{x,y}$.*

PROOF. The implications 1) $\Rightarrow$ 2) $\Rightarrow$ 3) are obvious. We prove that 3) $\Rightarrow$ 4). Suppose that all left cyclic polygons are weakly flat. It follows from the proof of Theorem 1 in [2] that S is then a regular monoid.

Let x and y be arbitrary elements of S. Then $S/\lambda_{x,y}$ is a left cyclic polygon with generator $[1]_{\lambda_{x,y}}$, which we denote by b, i.e., $S/\lambda_{x,y} = Sb$. By assumption the polygon Sb is weakly flat, and, since the equality

$$x \otimes b = 1 \otimes xb = 1 \otimes yb = y \otimes b$$

holds in $S \otimes Sb$, the equality $x \otimes b = y \otimes b$ must hold in $(xS \cup yS) \otimes Sb$. Hence, there must exist a finite chain

$$(x, b) = (x_1, b_1),\ (x_2, b_2), \dots, (x_k, b_k) = (y, b) \tag{5}$$

of pairs in $(xS \cup yS) \times Sb$ which joins (x, b) and (y, b). Let i be the smallest index such that $x_i \in yS$. (Such an i exists, since $x_k = y \in yS$.) We show that then $x_i \in xS \cap yS$. If $i = 1$, then $x_1 = x \in xS$, i.e., $x_1 \in xS \cap yS$. In the case $i \neq 1$ the pair (x_i, b_i) is obtained from (x_{i-1}, b_{i-1}) by transposition of an element in S, and $x_{i-1} \in xS$. It is easy to verify that this implies that $x_i \in xS \cap yS$. Since $b_i \in Sb$, it follows that $b_i = tb$ for some $t \in S$. It is clear that $x_i \otimes b_i = x \otimes b$, because the pair (x_i, b_i) belongs to the chain (5). By Lemma 2, the latter means that $xb = x_i b_i = x_i tb$. Let $z = x_i t$. Since $x_i \in xS \cap yS$, we have that $z \in xS \cap yS$ and $xb = zb$. The latter means that $(x, z) \in \lambda_{x,y}$. This proves 4).

We now prove the implication 4) $\Rightarrow$ 1). Let B be an arbitrary left S-polygon and R a right ideal of the monoid S, and assume that

$$x \otimes b_1 = y \otimes b_2 \tag{6}$$

in $S \otimes B$ for some x, $y \in R$ and b_1, $b_2 \in B$. Condition 1) will be proved if we show that (6) holds also in $R \otimes B$.

By Lemma 2, (6) gives us that $xb_1 = yb_2$, i.e., (6) can be realized by a chain with two transpositions

$$(x, b_1) \sim (1, xb_1) = (1, yb_2) \sim (y, b_2).$$

We now show that Lemma 3 can be applied. Indeed, if $sb = tb$ for some s, $t \in S$ and some $b \in B$, then $(s, t) \in \lambda$, where λ is the left congruence generated by b, i.e., $(s, t) \in \lambda \Leftrightarrow sb = tb$. Consequently, $\lambda_{s,t} \leq \lambda$, and by condition 4) there is a $z \in sS \cap tS$ with $(z, s) \in \lambda_{s,t} \leq \lambda$, i.e., $zb = sb = tb$. Thus, the conditions in Lemma 3 hold, and hence (6) is valid in $(xS \cup yS) \otimes B$ and, all the more so, in $R \otimes B$. The theorem is proved.

Let B be a left polygon and let $A \to C$ be an imbedding of right polygons. Let $W = W(A, C; B)$ denote the collection of pairs of elements in $A \times B$ such that $((a_1, b_1), (a_2, b_2)) \in W$ if and only if the equality $a_1 \otimes b_1 = a_2 \otimes b_2$ holds in $C \otimes B$ and not in $A \otimes B$. If B is not flat, then in general $W \neq \varnothing$. For each pair $((a_1, b_1), (a_2, b_2)) \in W$ we consider a chain of pairs in $C \times B$ joining (a_1, b_1) and (a_2, b_2) and having a minimal number of transpositions of elements in S. It is clear that among the pairs in W there is a pair $((a, b), (a'b')) \in W$ for which the length of such a chain is minimal. This chain is said to be an *absolutely shortest chain*, and is denoted by $Q(A, C; B)$.

LEMMA 5. *Let $Q = Q(A, C; B)$ be an absolutely shortest chain. Then the first transposition in it goes from left to right, each succeeding transposition goes in the direction opposite to the previous transposition, and the last transposition goes from right to left.*

PROOF. Suppose that Q has the form

$$(a, b) = (a_1, b_1),\ (a_2, b_2), \ldots, (a_k, b_k) = (a', b').$$

It is clear that transpositions in opposite directions alternate in Q, for two successive transpositions from left to right (from right to left) of elements $s \in S$ and $t \in S$ can be replaced by a single transposition of the element $ts \in S$ $(st \in S)$.

Assume that the first transposition in Q goes from right to left, i.e.,

$$(a, b) = (a_1, b_1) = (a_1, xb_2) \sim (a_1x, b_2) = (a_2, b_2).$$

Since $a_2 = a_1x \in A$ and the pairs (a_2, b_2) and (a', b') are joined by a chain shorter than Q, the equality $a_2 \otimes b_2 = a' \otimes b'$ holds in $A \otimes B$ because Q is absolutely minimal. But then, since $a \otimes b = a_2 \otimes b_2$ in $A \otimes B$, the equality $a \otimes b = a' \otimes b'$ holds in $A \otimes B$, and this contradicts the choice of Q. Thus, the first transposition in Q goes from left to right. It is proved similarly that the last transposition in Q goes from right to left, and the lemma follows.

Note that, by Lemma 5, the length of an absolutely shortest chain is always an even number.

Recall that a cyclic left S-polygon is called a *cyclic Rees polygon* if it is obtained from S by putting the elements of some left ideal of S together in a single class.

THEOREM 6. *For an arbitrary monoid S the following conditions are equivalent:*

1) *all cyclic Rees polygons are flat;*

2) *all cyclic Rees polygons are weakly flat;*

3) *S is a regular monoid, and the condition $(Sx \cup Sy) \cap xS \cap yS \neq \varnothing$ holds for any x, $y \in S$.*(1)

(1) *Added in proof.* This condition is equivalent to the condition $xS \cap yS \neq \varnothing$ for any monoid S.

PROOF. The implication 1) $\Rightarrow$ 2) is obvious. We prove that 2) $\Rightarrow$ 3). The regularity of S follows from the proof of Theorem 1 in [2], since the proof of this theorem actually uses only the weak flatness of cyclic Rees polygons. Let x and y be arbitrary elements in S, let $S/Sx \cup Sy$ be the cyclic Rees polygon with respect to the ideal $Sx \cup Sy$, and denote a generator of it by b, i.e., $S/Sx \cup Sy = Sb$. We consider the imbedding $xS \cup yS \to S$. Since

$$(x, b) = (1 \cdot x, b) \sim (1, xb) = (1, yb) \sim (y, b),$$

it follows that $x \otimes b = y \otimes b$ in $S \otimes Sb$. By assumption, the polygon Sb is weakly flat, i.e., the last equality must hold in the set $(xS \cup yS) \otimes Sb$. Consequently, there must exist a chain

$$(x, b) = (x_1, b_1), \dots, (x_k, b_k) = (y, b) \tag{7}$$

of pairs in $(xS \cup yS) \times Sb$ joining (x, b) and (y, b). Let i be a minimal index such that $x_i \in yS$. Note that then $x_i \in xS \cap yS$. Let $b_i = sb$. Since the pair (x_i, b_i) belongs to the chain (7), $x \otimes b = x_i \otimes sb$. By Lemma 2, the latter means that $xb = x_i sb$. Hence, $x_i s \in Sx \cup Sy$, and it follows from $x_i \in xS \cap yS$ that $x_i s \in xS \cap yS$. Thus,

$$x_i s \in (Sx \cup Sy) \cap xS \cap yS.$$

We now show that 3) $\Rightarrow$ 1). Let $Sb = S/V$ be an arbitrary cyclic Rees polygon with respect to some left ideal V. Let aS be an arbitrary cyclic right polygon, and A' a subpolygon of it. By Theorem 1, it suffices for us to show that the imbedding $A' \to aS$ induces an imbedding $A' \otimes Sb \to aS \otimes Sb$.

Assume not. Then $W(A', aS; Sb) \neq \varnothing$, and hence there exists an absolutely shortest chain $Q = Q(A', aS; Sb)$. Suppose that it joins the pairs (ax, sb) and (ay, tb) and has the form

$$(ax, sb) = (ax_1, s_1 b),\ (ax_2, s_2 b), \dots, (ax_k, s_k b) = (ay, tb). \tag{8}$$

Suppose that the pair $(ax_i, s_i b)$, $1 < i \leq k$, is obtained from the pair $(ax_{i-1}, s_{i-1} b)$ by transposition of an element $y_{i-1} \in S$. Assume that the chain (8) contains at least four transpositions. By Lemma 5, the first transposition of y_1 is from left to right, i.e., (8) has the following form:

$$\begin{aligned}(ax, sb) &= (ax_1, s_1 b) = (ax_2 y_1, s_1 b) \sim (ax_2, y_1 s_1 b)\\ &= (ax_2, s_2 b) = (ax_2, y_2 s_3 b) \sim (ax_2 y_2, s_3 b)\\ &= (ax_3, s_3 b) = (ax_4 y_3, s_3 b) \sim (ax_4, y_3 s_3 b)\\ &= (ax_4, s_4 b) = (ax_4, y_4 s_5 b) \sim (ax_4 y_4, s_5 b)\\ &= (ax_5, s_5 b) \sim \cdots \sim (ay, tb).\end{aligned} \tag{9}$$

This implies that $y_1 s_1 b = s_2 b = y_2 s_3 b$ and $y_3 s_3 b = s_4 b = y_4 s_5 b$. Since Sb is the cyclic Rees polygon S/V, this means that either $y_1 s_1 = y_2 s_3$ or $y_1 s_1$, $y_2 s_3 \in V$, and, similarly, either $y_3 s_3 = y_4 s_5$ or $y_3 s_3$, $y_4 s_5 \in V$.

I. Assume that y_1s_1, y_2s_3, y_3s_3, $y_4s_5 \in V$. Then $x_2y_1s_1 \in V$ and $x_4y_4s_5 \in V$, i.e., $x_2y_1s_1b = x_4y_4s_5b$. This shows us that the first four transpositions of the chain (9) can be replaced by the two transpositions

$$\begin{aligned}(ax, sb) &= (ax_1, s_1b) = (ax_2y_1, s_1b) \sim (a, x_2y_1s_1b) \\ &= (a, x_4y_4s_5b) \sim (ax_4y_4, s_5b) = (ax_5, s_5b).\end{aligned}$$

II. Assume that $y_1s_1 = y_2s_3$ and $y_3s_3 = y_4s_5$. Since $ax_2y_2 = ax_3 = ax_4y_3$, we then get that

$$\begin{aligned}(ax, sb) &= (ax_1, s_1b) = (ax_2y_1, s_1b) \sim (ax_2y_1s_1, b) \\ &= (ax_2y_2s_3, b) = (ax_4y_3s_3, b) = (ax_4y_4s_5, b) \\ &\sim (ax_4y_4, s_5b) = (ax_5, s_5b),\end{aligned}$$

i.e., the first four transpositions of the chain (9) were replaced by two transpositions.

III. Finally, assume that $y_1s_1 = y_2s_3$, while y_3s_3, $y_4s_5 \in V$, i.e., $y_3s_3b = y_4s_5b$. (The case y_1s_1, $y_2s_3 \in V$, $y_3s_3 = y_4s_5$ is analogous.) This time the first four transpositions of the chain (9) can be replaced by the following three transpositions:

$$\begin{aligned}(ax, sb) &= (ax_1, s_1b) = (ax_2y_1, s_1b) \sim (ax_2y_1s_1, b) \\ &= (ax_2y_2s_3, b) = (ax_4y_3s_3, b) \sim (ax_4, y_3s_3b) \\ &= (ax_4, y_4s_5b) \sim (ax_4y_4, s_5b) = (ax_5, s_5b).\end{aligned}$$

Thus, it is shown that in all the cases considered the absolutely shortest chain Q can be made shorter. Since an absolutely shortest chain contains an even number of transpositions, this contradiction proves that Q contains at most two transpositions, i.e., it has the form

$$\begin{aligned}(ax, sb) &= (ax_1, s_1b) = (ax_2y_1, s_1b) \sim (ax_2, y_1s_1b) \\ &= (ax_2, s_2b) = (ax_2, y_2s_3b) \sim (ax_2y_2, s_3b) = (ay, tb). \qquad (10)\end{aligned}$$

From this, $y_1s_1b = s_2b = y_2s_3b$, which means that either $y_1s_1 = y_2s_3$ or y_1s_1, $y_2s_3 \in V$. If $y_1s_1 = y_2s_3$, then the chain (10) can be replaced by the chain

$$\begin{aligned}(ax, sb) &= (ax_2y_1, s_1b) \sim (ax_2y_1s_1, b) = (ax_2y_2s_3, b) \\ &\sim (ax_2y_2, s_3b) = (ay, tb)\end{aligned}$$

of pairs in $A' \otimes Sb$. If y_1s_1, $y_2s_3 \in V$, then we use the assumption that $(Sy_1s_1 \cup Sy_2s_3) \cap y_1s_1S \cap y_2s_3S \neq \varnothing$. Suppose that z belongs to this intersection, i.e., $z = y_1s_1v_1 = y_2s_3v_2 \in V$. Then $zb = y_1s_1b = y_2s_3b$. We now show that the chain (10) can be replaced by a chain of pairs in $A' \otimes Sb$. Indeed, the chain

$$\begin{aligned}(ax, sb) &= (ax_2y_1, s_1b) = (ax_2y_1y_1^*y_1, s_1b) \sim (ax_2y_1y_1^*, y_1s_1b) \\ &= (ax_2y_1y_1^*, y_1s_1v_1b) \sim (ax_2y_1y_1^*y_1s_1v_1, b) = (ax_2y_1s_1v_1, b) \\ &= (ax_2y_2s_3v_2, b) = (ax_2y_2y_2^*y_2s_3v_2, b) \sim (ax_2y_2y_2^*, y_2s_3v_2b) \\ &= (ax_2y_2y_2^*, y_2s_3b) \sim (ax_2y_2y_2^*y_2, s_3b) = (ax_2y_2, s_3b) = (ay, tb)\end{aligned}$$

shows that $ax \otimes sb = ay \otimes tb$ in $A' \otimes Sb$; but the equality $ax \otimes sb = ay \otimes tb$ must not hold in $A' \otimes Sb$ by the definition of Q. This contradiction proves that the shortest chain Q does not exist, i.e., $W(A', aS, Sb) = \varnothing$. Consequently, Sb is a flat polygon. The theorem is proved.

It follows from Theorem 6 that, in particular, all cyclic Rees polygons over a regular monoid with zero are flat.

THEOREM 7. *For an arbitrary monoid S the following conditions are equivalent:*

1) *S is a completely flat monoid;*
2) *all right and left finitely generated S-polygons are flat;*
3) *all right and left cyclic S-polygons are flat;*
4) *all right and left S-polygons are weakly flat;*
5) *all right and left finitely generated S-polygons are weakly flat;*
6) *all right and left cyclic S-polygons are weakly flat;*
7) *S is a regular monoid, and for arbitrary x, $y \in S$ there exist elements $u \in xS \cap yS$, $v \in Sx \cap Sy$ such that $(u, x) \in \lambda_{x,y}$ and $(v, x) \in \rho_{x,y}$.*

PROOF. The implications 1) $\Rightarrow$ 2) $\Rightarrow$ 3) $\Rightarrow$ 6) and 1) $\Rightarrow$ 4) $\Rightarrow$ 5) $\Rightarrow$ 6) arc obvious. The implication 6) $\Rightarrow$ 7) follows from Theorem 4 and its right-sided version.

We prove that 7) $\Rightarrow$ 1). Let B be an arbitrary left polygon, $A = aS$ an arbitrary right cyclic polygon, and A' a subpolygon of it. Obviously, it suffices for us to prove that $W(A', A; B) = \varnothing$. Assume the opposite; then there exists an absolutely shortest chain $Q = Q(A', A; B)$ joining the pairs (as, b) and (at, b'), where as, $at \in A'$ and b, $b' \in B$. This means that the equality

$$as \otimes b = at \otimes b' \tag{11}$$

holds in $aS \otimes B$, but not in $A' \otimes B$. Suppose that Q has the form

$$(as, b) = (as_1, b_1),\ (as_2, b_2), \dots, (as_m, b_m) = (at, b'). \tag{12}$$

By Lemma 5, the first transposition in (12) is from left to right, and the length of the chain is an even number $2k$. Assume that $k \geq 2$, i.e., the chain (12) has the form

$$\begin{aligned}(as, b) &= (as_1, b_1) = (as_2x_1, b_1) \sim (as_2, x_1b_1)\\ &= (as_2, b_2) = (as_2, x_2b_3) \sim (as_2x_2, b_3)\\ &= (as_3, b_3) = (as_4x_3, b_3) \sim (as_4, x_3b_3)\\ &= (as_4, b_4) = (as_4, x_4b_5) \sim (as_4x_4, b_5)\\ &= (as_5, b_5) \sim \cdots \sim (at, b').\end{aligned} \tag{13}$$

We note that from this chain we can get a new chain joining (as, b) and (at, b'), in which after each transposition from left to right the element a is in the first position of the corresponding pair. The new chain has the form

$$\begin{aligned}(as, b) &= (as_1, b_1) = (as_2x_1, b_1) \sim (a, s_2x_1b_1)\\ &= (a, s_2b_2) = (a, s_2x_2b_3) \sim (as_2x_2, b_3)\\ &= (as_3, b_3) = (as_4x_3, b_3) \sim (a, s_4x_3b_3)\\ &= (a, s_4b_4) = (as_4, b_4) \sim \cdots \sim (at, b').\end{aligned}$$

Thus, we have obtained a chain of transpositions of the type

$$\begin{aligned}(as, b) &= (ay_1, b_1) \sim (a, y_1b_1) = (a, y_2b_2) \sim (ay_2, b_2)\\ &= (ay_3, b_2) \sim (a, y_3b_2) = (a, y_4b_3)\\ &\sim (ay_4, b_3) = (ay_5, b_3) \sim \cdots \sim (at, b'). \qquad (14)\end{aligned}$$

This gives us that $y_1b_1 = y_2b_2 = c \in B$, $y_3b_2 = y_4b_3 = d \in B$, and $ay_2 = ay_3$. Consequently, $y_1y_1^*c = y_2y_2^*c$, $y_3y_3^*d = y_4y_4^*d$, and $ay_2 = ay_3$. By assumption, this implies the existence of elements

$$e = y_1y_1^*u_1 = y_2y_2^*u_2, \qquad (15)$$
$$f = y_3y_3^*u_3 = y_4y_4^*u_4, \qquad (16)$$
$$g = vy_2 = wy_3 \qquad (17)$$

such that

$$ec = (y_1y_1^*u_1)c = (y_2y_2^*u_2)c = (y_1y_1^*)c = (y_2y_2^*)c, \qquad (18)$$
$$fd = (y_3y_3^*u_3)d = (y_4y_4^*u_4)d = (y_3y_3^*)d = (y_4y_4^*)d, \qquad (19)$$
$$ag = a(vy_2) = a(wy_3) = ay_2 = ay_3. \qquad (20)$$

We now have the following chain:([2])

$$\begin{aligned}(ay_1y_1^*u_1, c) &=^{(15)} (ay_2y_2^*u_2, c) =^{(20)} (avy_2y_2^*u_2, c)\\ &\sim (a, vy_2y_2^*u_2c) =^{(18)} (a, vy_2y_2^*c) = (a, vy_2y_2^*y_2b_2)\\ &= (a, vy_2b_2) =^{(17)} (a, wy_3b_2) = (a, wy_3y_3^*y_3b_2) = (a, wy_3y_3^*d)\\ &=^{(19)} (a, wy_3y_3^*u_3d) \sim (awy_3y_3^*u_3, d)\\ &=^{(20)} (ay_3y_3^*u_3, d) =^{(16)} (ay_4y_4^*u_4, d),\end{aligned}$$

which contains two transpositions. Thus, we have proved that from a chain containing four transpositions and joining (ay_1, b_1) and (ay_4, b_3) we can get a chain consisting of two transpositions and joining $(ay_1y_1^*u_1, c)$ and $(ay_4y_4^*u_4, d)$, and $y_1y_1^*u_1c = y_1y_1^*c = y_1y_1^*y_1v_1 = y_1b_1$; similarly, $y_4y_4^*u_4d = y_4b_3$.

Assume now that from the subchain L of (14) containing the first $2l$ ($l < k$) transpositions and joining (ay_1, b_1) and (ay_i, b_i) we can get a chain L' of $2l-2$ transpositions joining (ay_1r_1, t_1b_1) and (ay_ir_i, t_ib_i), and $y_1r_1t_1b_1 = y_1b_1$ and

([2]) The numbers after equalities in the chain denote the labels of the relations used in the corresponding step.

$y_ir_it_ib_i = y_ib_i$. We show that the analogous result holds for the subchain of (14) containing the first $2l+2$ transpositions. Indeed, suppose that in (14)

$$(ay_i, b_i) = (ay_{i+1}, b_i) \sim (a, y_{i+1}b_i) = (a, y_{i+2}b_{i+1}) \sim (ay_{i+2}, b_{i+1}),$$

i.e., $y_{i+1}b_i = y_{i+2}b_{i+1} = q$, and hence $y_{i+1}y^*_{i+1}q = y_{i+2}y^*_{i+2}q$ and $ay_i = ay_{i+1}$. By condition 7), this implies that there exist elements

$$y_{i+1}y^*_{i+1}z_1 = y_{i+2}y^*_{i+2}z_2, \tag{21}$$

$$xy_i = zy_{i+1} \tag{22}$$

such that

$$y_{i+1}y^*_{i+1}q = y_{i+1}y^*_{i+1}z_1q = y_{i+2}y^*_{i+2}z_2q = y_{i+2}y^*_{i+2}q, \tag{23}$$

$$axy_i = ay_i = ay_{i+1} = azy_{i+1}. \tag{24}$$

We now get that

$$\begin{aligned}(ay_ir_i, t_ib_i) &= {}^{(24)}(axy_ir_i, t_ib_i) \sim (a, xy_ir_it_ib_i) = (a, xy_ib_i)\\ &= {}^{(22)}(a, zy_{i+1}b_i) = (a, zy_{i+1}y^*_{i+1}y_{i+1}b_i) = (a, zy_{i+1}y^*_{i+1}q)\\ &= {}^{(23)}(a, zy_{i+1}y^*_{i+1}z_1q) \sim (azy_{i+1}y^*_{i+1}z_1, q) =^{(24)} (ay_{i+1}y^*_{i+1}z_1, q)\\ &= {}^{(21)}(ay_{i+2}y^*_{i+2}z_2, q) = (ay_{i+2}y^*_{i+2}z_2, y_{i+2}b_{i+1}).\end{aligned}$$

This chain contains two transpositions, and by supplementing the chain L' with it we get a chain L'' containing $2l$ transpositions and joining (ay_1r_1, t_1b_1) and $(ay_{i+2}y^*_{i+2}z_2, y_{i+2}b_{i+1})$, and

$$y_{i+2}y^*_{i+2}z_2y_{i+2}b_{i+1} =^{(23)} y_{i+2}y^*_{i+2}y_{i+2}b_{i+1} = y_{i+2}b_{i+1}.$$

Thus, this implies that from a chain Q joining (as, b) and (at, b') and containing $2k$ transpositions we can obtain a chain Q' containing $2k-2$ transpositions and joining (asr_1, t_1b) and (atr_2, t_2b'), and we also have $sr_1t_1b = sb$ and $tr_2t_2b' = tb'$. Since Q is an absolutely shortest chain and $asr_1, atr_2 \in A'$, the equality $asr_1 \otimes t_1b = atr_2 \otimes t_2b'$ holds in $A' \otimes B$. On the other hand, the existence of the chain

$$\begin{aligned}(as, b) &= (ass^*s, b) \sim (ass^*, sb) = (ass^*, sr_1t_1b)\\ &\sim (ass^*sr_1, t_1b) = (asr_1, t_1b)\end{aligned}$$

means that $as\otimes b = asr_1 \otimes t_1b$ in $A'\otimes B$. Similarly, $at\otimes b' = atr_2\otimes t_2b'$ in $A'\otimes B$. Consequently, $as \otimes b = at \otimes b'$ in $A' \otimes B$, which contradicts the existence of an absolutely shortest chain Q joining (as, b) and (at, b'). We have obtained a contradiction by assuming that Q contains at least four transpositions ($k \geq 2$). Consequently, the chain contains only two transpositions, and hence the conditions of Lemma 3 are satisfied. Using this lemma, we get that the equality $as \otimes b = at \otimes b'$ holds in $A' \otimes B$. We have again obtained a contradiction, which means that $W(A', aS, B) = \varnothing$, i.e., B is a flat polygon.

Since condition 7) is symmetric, it can be proved similarly that any right polygon is flat. Thus, the implication 7) $\Rightarrow$ 1) is valid, and the theorem is proved.

COROLLARY 8. *Let S be a monoid in which the idempotents commute. All left (right) S-polygons are flat if and only if S is an inverse monoid.*

PROOF. *Necessity.* It follows from Theorem 4 that S is a regular monoid and, consequently ([6], Theorem 7.4), S is an inverse monoid.

Sufficiency. Let S be an inverse monoid. We show that condition 7) of Theorem 7 holds. Let x and y be arbitrary elements of S. Then it follows from $(x, y) \in \lambda_{x,y}$ that $(xx^*x, xx^*y) \in \lambda_{x,y}$, i.e., $(x, xx^*yy^*y) \in \lambda_{x,y}$. But $xx^*yy^*y = yy^*xx^*y \in xS \cap yS$. Similarly, $(x, yy^*yx^*x) \in \rho_{x,y}$, and $yy^*yx^*x = yx^*xy^*y \in Sx \cap Sy$. By Theorem 7, this implies that all left (and right) S-polygons are flat.

In conclusion the author thanks M. Kil'p for useful discussions.

BIBLIOGRAPHY

1. M. P. Dorofeeva, *Some properties of the category of polygons*, Sibirsk. Mat. Zh. **14** (1973), 906; English transl. in Siberian Math. J. **14** (1973).

2. M. Kil'p, *On flat polygons*, Tartu Riikl. Ül. Toimetised No. 253 (1970), 66–72. (Russian)

3. ——, *On the homological classification of monoids*, Sibirsk. Mat. Zh. **13** (1972), 578–586; English transl. in Siberian Math. J. **13** (1972).

4. ——, *On completely flat monoids*, Second All-Union Sympos. Theory of Semigroups, Abstracts of Reports, Ural. Gos. Univ., Sverdlovsk, 1978, p. 33. (Russian)

5. Joachim Lambek, *Lectures on rings and modules*, Blaisdell, Waltham, Mass., 1966.

6. E. S. Lyapin, *Semigroups*, Fizmatgiz, Moscow, 1960; English transl., Amer. Math. Soc., Providence, R. I., 1963.

7. L. A. Skornyakov, *On the homological classification of monoids*, Sibirsk. Mat. Zh. **10** (1969), 1139–1143; English transl. in Siberian Math. J. **10** (1969).

8. Bo Stenström, *Flatness and localization over monoids*, Math. Nachr. **48** (1971), 315–334.

Translated by H. H. MCFADEN

Amer. Math. Soc. Transl.
(2) Vol. **142**, 1989

On a Special Representation of Polynomials that are Positive on a System of Closed Intervals

UDC 517.5

M. G. KREĬN, B. YA. LEVIN, A. A. NUDEL'MAN

Abstract. Conditions are found for a polynomial which is positive on a system of closed intervals to admit a special representation analogous to that found by A. A. Markov for a polynomial which is positive on a closed interval. A construction of the solution is given when these conditions are satisfied. Moreover, a solution is obtained for the problem of a polynomial that has the least deviation from zero on a system of intervals, with a special weight. The solution uses methods and problems from the Nevanlinna–Pick interpolation theory, the Stieltjes moment problem, and the theory of Abelian integrals, as well as properties of analytic functions that belong to special classes.

The solutions of these problems are applied to the study of the spectral functions of a string.

Bibliography: 12 titles.

Introduction

This investigation arose as a result of our study and discussion of a paper by Akhiezer [1] in which he solved the inverse Sturm–Liouville problem for a class of spectral functions, and incidentally obtained a representation for polynomials that are positive on a system of intervals $[0, k^2]$ and $[1, \infty)$ $(0 < k^2 < 1)$.

Under certain conditions this representation can be considered as a generalization of the following result of A. A. Markov (1906; see [2]).

THEOREM (MARKOV). *Every polynomial $P(z)$ that is of even degree $N = 2n$ and positive on the interval $[a, b]$ has the representation*

$$P(z) = A^2(z) + (z-a)(b-z)B^2(z), \tag{0.1}$$

1980 *Mathematics Subject Classification* (1985 *Revision*). Primary 26C99, 41A10, 41A50; Secondary 70J10.

Translation of Preprint No. 28–84, Fiz.-Tekhn. Inst. Nizkikh Temperatur Akad. Nauk Ukrain. SSR, Kharkov, 1984.

where $A(z)$ and $B(z)$ are real polynomials of respective degrees n and $n-1$, with simple zeros that are on $[a, b]$ and interlaced. Under these conditions $A(z)$ and $B(z)$ are uniquely determined up to sign.

The Fejér–Riesz theorem (1916) on the representation of trigonometric polynomials that are positive on $[0, 2\pi]$ can be considered as a generalization of this result.

Markov also raised the question of constructing $A(z)$ and $B(z)$ in terms of the zeros of $P(z)$.

An interval $[a, b]$ is formed by deleting from the real line the intervals $(-\infty, a)$ and (b, ∞). Let us denote by E a set obtained by deleting from the real line a finite collection of disjoint intervals (some of which may be unbounded). With each interval (α, β) we associate the simplest polynomial that is positive outside $[\alpha, \beta]$ and negative in (α, β); specifically, $(z-\alpha)(z-\beta)$ for $-\infty < \alpha < \beta < \infty$, $z - \beta$ for $\alpha = -\infty$, and $\alpha - z$ for $\beta = \infty$. Let $T_E(z)$ denote the product of all the simplest polynomials corresponding to the deleted intervals. For a polynomial $P(z)$ which is positive on E an analog of the representation (0.1) is the representation

$$P(z) = A^2(z) + T_E(z)B^2(z), \tag{0.2}$$

in which the zeros of the real polynomials $A(z)$ and $B(z)$ are simple, belong to E, and are interlaced in a particular way on the intervals that make up E (for more details see §1). A pair of polynomials $A(z)$, $B(z)$ that have these properties will be called a *special pair*, and the representation (0.2) will be called a *special representation* of $P(z)$. In contrast to Markov's case, a representation (0.2) is not always possible, and we have the following problem:

PROBLEM A. *Find necessary and sufficient conditions for a polynoial $P(z)$, positive on E, to have a special representation.*

Except in Markov's case, there is always an interval (α, β) adjacent to E. From a polynomial $P(z)$ which is positive on E we can always pass to a polynomial $(z-1)^N P((\alpha z - \beta)/(z-1))$ which is positive on a system E_0 of intervals lying on the positive real line. Under such a transformation, $T_E(z)$ is carried to a polynomial which differs from $T_{E_0}(z)$ only by a positive constant multiple. For a number of reasons, both of a theoretical and a technical nature, it is precisely this case that warrants detailed study. Hence we shall investigate the following problem.

PROBLEM A_0. *Find necessary and sufficient conditions for there to exist a special representation of a polynomial $P(z)$ which is positive on $E_0 = [0, \infty) \setminus \bigcup_1^g (a_j, b_j)$, where $0 < a_1 < b_1 < \cdots < a_g < b_g$.*

In solving this problem we may omit the requirement that $P(z)$ is of even degree.

Markov used the representation (0.1) to solve the problem of the least deviation from zero, on $[a, b]$, of functions of the form $X(z)/\sqrt{P(z)}$, where $X(z)$

is a polynomial of degree n with leading coefficient 1, and $P(z)$ is a given polynomial of degree $2n$, positive on $[a, b]$. Apparently the analogous problem for a system of finite intervals has not previously been solved. However, as it turns out, it can be solved in the case when $P(z)$ has a special representation. Here we establish that the minimizing polynomial differs from $A(z)$ only by a constant multiplier. Moreover, this suggests a second problem on least deviation, in which the minimizing polynomial differs from $B(z)$ by a constant multiplier.

To solve these problems on the special representation we use methods and results from the theory of interpolation problems of Nevanlinna–Pick type, from the theory of the Stieltjes moment problem, for the theory of Abelian integrals, and also the properties of certain classes of analytic functions.

We obtain the solution of Problem A_0 in two forms which do not directly reduce to each other and which are in some sense complementary. We also obtain two forms of an expression which lets us construct the polynomials $A(z)$ and $B(z)$ from the zeros of $P(z)$. One of these has let us connect Problem A_0 with a problem from the theory of the spectral functions of a string. This problem is also connected with a result of Akhiezer, which we generalize.

§1. Special pairs of polynomials and some classes of analytic functions

1. Unless the contrary is stated, we shall take E to be E_0, that is, a set $[0, \infty)\setminus\bigcup_1^g(a_j, b_j)$, where $0 < a_1 < b_1 < \cdots < a_g < b_g < \infty$. It will be convenient to set $a(z) = \prod_1^g(z - a_j)$ and $b(z) = \prod_1^g(z - b_j)$. Then $T(z) = T_{E_0}(z) = za(z)b(z)$.

We say that a pair $A(z)$, $B(z)$ of polynomials is *special* (with respect to E_0) if the following conditions are satisfied:

a) $A(z)$ and $B(z)$ are real polynomials;

b) $A(z)$ and $B(z)$ have respective degrees $[N/2]$ and $[(N-1)/2] - g$ for some positive integer $N \geq 2g + 1$;

c) all the zeros of $A(z)$ and $B(z)$ are simple and belong to E_0;

d) in each finite interval (b_{j-1}, a_j), $j = 1, 2, \ldots, g$ $(b_0 = 0)$, $A(z)$ has one more zero than $B(z)$; in the intervals (b_{j-1}, a_j), $j = 1, 2, \ldots, g+1$ $(a_{g+1} = \infty)$ there is precisely one zero of $B(z)$ between each two consecutive zeros of $A(z)$. In other words:

d′) the zeros of $A(z)$ and $zb(z)B(z)$ are interlaced;

e) $A(z)$ and $zB(z)$ have the same sign in $(-\infty, 0)$ and consequently in (a_j, b_j), $j = 1, 2, \ldots, g$.

The definition of a special pair can be formulated in terms of its function-theoretic properties. For this purpose we introduce some standard classes of analytic functions, with the notation used in [3].

2. A function $F(z)$ belongs to class $\mathscr{R}$ if it is holomorphic in the upper half plane and has nonnegative imaginary part there. We know (see, for example, [3]) that $F(z) \in \mathscr{R}$ if and only if it has the representation

$$F(z) = \alpha + \beta z + \int_{-\infty}^{\infty} \left(\frac{1}{t - z} - \frac{t}{1 + t^2} \right) d\sigma(t),$$

where $\overline{\alpha} = \alpha$, $\beta \geq 0$, $d\sigma(t) \geq 0$, $\int_{-\infty}^{\infty} (1 + t^2)^{-1} d\sigma(t) < \infty$.

The constants α and β and the measure $d\sigma(t)$ are uniquely determined by $F(z)$: if t_1 and t_2 are points of continuity of $\sigma(t)$, then

$$\sigma(t_2) - \sigma(t_1) = \frac{1}{\pi} \lim_{\varepsilon \downarrow 0} \int_{t_1}^{t_2} \operatorname{Im} F(x + i\varepsilon)\, dx. \tag{1.1}$$

Hence it follows, in particular, that if $F(z)$ is holomorphic and real in an interval of the real axis, then $d\sigma(t) \equiv 0$ in that interval.

If $F(z) \not\equiv 0$, then the relations $F(z) \in \mathscr{R}$ and $-1/F(z) \in \mathscr{R}$ are equivalent. From $F(z) \in \mathscr{R}$ and $G(z) \in \mathscr{R}$ it follows that $\sqrt{F(z)G(z)} \in \mathscr{R}$.

Let S denote the class of functions in $\mathscr{R}$ that admit a holomorphic continuation to $\mathbb{C} \backslash [0, \infty)$ and are nonnegative on $(-\infty, 0)$. Functions of this class are characterized by the representation

$$F(z) = \gamma + \int_0^{\infty} \frac{d\sigma(t)}{t - z} \qquad (\gamma = F(-\infty) \geq 0,\ d\sigma(t) \geq 0).$$

When $F(z) \not\equiv 0$ the relations $F(z) \in S$ and $-1/(zF(z)) \in S$ are equivalent.

For a set $E = E_0$ $(= [0, \infty) \backslash \bigcup_1^g (a_j, b_j))$ we denote by $S(E)$ the class of functions in S which are holomorphic and real on $[0, \infty) \backslash E$. A function $F(z)$ belongs to this class if and only if

$$F(z) = \gamma + \int_E \frac{d\sigma(t)}{t - z} \qquad (\gamma = F(-\infty) \geq 0,\ d\sigma(t) \geq 0).$$

Finally, we denote by $S_+(E)$ the class of functions in S which are holomorphic and nonnegative in $[0, \infty) \backslash E$.

It is known [3](1) that $F(z) \in S_+(E)$ if and only if $F(z) \in \mathscr{R}$ and $(zb(z)/a(z))F(z) \in \mathscr{R}$ and also if and only if $F(z) \in S$ and $(b(z)/a(z))F(z) \in S$. Evidently $S_+(E) \subset S(E) \subset S \subset \mathscr{R}$.

3. LEMMA 1.1. *For polynomials $A(z)$ and $B(z)$ that satisfy conditions* a) *and* b), *the following statements are equivalent (up to a possible change in the sign of $B(z)$):*

1) *$A(z)$ and $B(z)$ form a special pair;*
2) $A(z)/(zb(z)B(z)) \in S_+(E)$;
3) $\operatorname{Im}(\sqrt{T(z)}B(z)/A(z)) > 0$ *for* $z \in \mathbb{C} \backslash E$; (1.2)
4) $\left| \dfrac{A(z) + i\sqrt{T(z)}B(z)}{A(z) - i\sqrt{T(z)}B(z)} \right| < 1$ *for* $z \in \mathbb{C} \backslash E$.

(1)In [3] the class $S_+(E)$ is denoted by $S(E_g)$.

Here and later we understand by $\sqrt{T(z)}$ the branch in $\mathbb{C}\backslash E$ which is positive on the upper side of the cut $(0, a_1)$.

PROOF. Let $A(z)$ and $B(z)$ form a special pair. By conditions c) and d), when $A(z)$ and $B(z)$ are a special pair the function $F(z) = A(z)/(zb(z)B(z))$ is holomorphic and positive in $(-\infty, \infty)\backslash E$. We verify that $F(z) \in \mathscr{R}$, whence the implication 1) $\Rightarrow$ 2) follows. Since the zeros of $A(z)$ and $zb(z)B(z)$ are interlaced, and $F(z)$ is positive on $(-\infty, 0)$, it follows that the residues of $F(z)$ are negative, that is,

$$F(z) = \frac{A(z)}{zb(z)B(z)} = \gamma + \sum_{j=1}^{[(N+1)/2]} \frac{\mu_j}{\xi_j - z} \qquad (\gamma \geq 0,\ \mu_j > 0). \tag{1.3}$$

Consequently $\operatorname{Im} F(z) > 0$ for $\operatorname{Im} z > 0$, and $F(z) \in \mathscr{R}$.

To prove that 2) $\Rightarrow$ 1), we observe that since the rational function $F(z) = A(z)/(zb(z)B(z))$ belongs to $S_+(E)$, it is representable in the form (1.3), where the ξ_j belong to E. It follows from this representation that its poles, and likewise its zeros, are real and simple; also the zeros and poles are interlaced, and the zeros also belong to E.

We now prove that 2) $\Rightarrow$ 3). If $F(z) = A(z)/(zb(z)B(z)) \in S_+(E)$ then $F(z) \in \mathscr{R}$ and

$$G(z) = \left(\frac{zb(z)}{a(z)}\right) F(z) = \frac{A(z)}{a(z)B(z)} \in \mathscr{R}.$$

Therefore

$$\sqrt{T(z)}B(z)/A(z) = \sqrt{(-1/F(z))(-1/G(z))} \in \mathscr{R}.$$

Consequently $\operatorname{Im} \sqrt{T(z)}B(z)/A(z) > 0$ for $\operatorname{Im} z > 0$. It follows by continuity that this inequality is preserved for $z \in (-\infty, \infty)\backslash E$. We next prove that it also holds for $\operatorname{Im} z < 0$. Let $\zeta = \overline{z}$, $\operatorname{Im} \zeta > 0$. Then, since $A(z)$ and $B(z)$ are real,

$$\arg(\sqrt{T(z)}B(z)/A(z)) = \pi - \arg(\sqrt{T(\zeta)}B(\zeta)/A(\zeta)),$$

whence

$$\operatorname{Im}(\sqrt{T(z)}B(z)/A(z)) = \operatorname{Im}(\sqrt{T(\zeta)}B(\zeta)/A(\zeta)) > 0.$$

Conversely, let 3) be true; since the residues of $a(z)/(zb(z))$ are positive, we have $\operatorname{Im}(a(z)/(zb(z))) < 0$ for $\operatorname{Im} z > 0$, so that

$$\pi/2 < \arg \sqrt{a(z)/(zb(z))} < \pi. \tag{1.4}$$

In addition, since $\sqrt{T(z)}B(z)/A(z) = \sqrt{a(z)/(zb(z))} \cdot zb(z)B(z)/A(z)$, we find from (1.2) that

$$0 < \arg(\sqrt{a(z)/(zb(z))} \cdot zb(z)B(z)/A(z)) < \pi,$$

whence, by using (1.4), we obtain

$$2\pi < \arg \sqrt{a(z)/(zb(z))} + \arg(zb(z)B(z)/A(z)) < 3\pi.$$

Consequently,

$$(2\pi >) \arg(zb(z)B(z)/A(z)) > 2\pi - \arg\sqrt{a(z)/(zb(z))} > \pi,$$

that is, $\operatorname{Im}(A(z)/(zb(z)B(z))) > 0$ when $\operatorname{Im} z > 0$; in other words, $A(z)/(zb(z)B(z)) \in \mathscr{R}$. Finally, since when $z \in (-\infty, \infty)\backslash E$ we have $\sqrt{a(z)/(zb(z))} > 0$, $\operatorname{Im}\sqrt{T(z)}B(z)A(z) > 0$ and $\operatorname{Im}(zb(z)B(z)/A(z)) = 0$, we have $A(z)/(zb(z)B(z)) > 0$ when $z \in (-\infty, \infty)\backslash E$.

The equivalence of 3) and 4) is evident. This completes the proof of the lemma.

§2. On an interpolation problem and the related moment problem

The solution of the problem of the existence of a special pair $A(z)$, $B(z)$ with the representation (0.2) is closely connected with an interpolation problem of Nevanlinna–Pick type, and via that to the power-moment problem on E.

Let Z be the set of zeros of the polynomial $P(z)$, positive on E, that lie in the upper half plane and on $(-\infty, \infty)\backslash E$; the zeros are counted according to their multiplicities. Also let a function $w(z) \in S(E)$ be given.

We denote by $\mathrm{NP}(Z, w)$ the problem of finding all functions $F \in S(E)$ for which every point of Z is a zero of the difference $F(z) - w(z)$ (and of its derivatives of the corresponding orders in the multiple case).

We denote by s_k ($k = 0, 1, \dots, N-1$) a number determined in the following way. We decompose the quotient $t^k/P(t)$ into a sum of partial fractions and supply to this decomposition the linear functional that assigns to each fraction of the form $1/(t - z_0)^{l+1}$ the number $\overline{(w(z) - w(-\infty))^{(l)}|_{z=z_0}/l!}$ if $\operatorname{Im} z_0 \geq 0$ and the number $(w(z) - w(-\infty))^{(l)}|_{z=\bar{z}_0}/l!$ if $\operatorname{Im} z < 0$. We denote by $\mathrm{M}(Z, w)$ the moment problem that consists of finding the measures $d\tau(t)$ and numbers $\rho \geq 0$ that provide the representation

$$\begin{aligned} s_k &= \int_E t^k\, d\tau(t), \quad k = 0, 1 \dots, N-2, \\ s_{N-1} &= \int_E t^{N-1}\, d\tau(t) + \rho. \end{aligned} \tag{2.1}$$

LEMMA 2.1. *If*

$$F(z) = \gamma + \int_E \frac{d\sigma(t)}{t - z} \quad (\gamma \geq 0,\ d\sigma(t) \geq 0) \tag{2.2}$$

is a solution of the problem $\mathrm{NP}(Z, w)$, *then the measure* $d\tau(t) = d\sigma(t)/P(t)$ *and* $\rho = \gamma/P_0$ (P_0 *is the leading coefficient of* $P(t)$) *form a solution of the moment problem* $\mathrm{M}(Z, w)$. *Conversely, if the measure* $d\tau(t)$ *and the number* $\rho \geq 0$ *form a soluton of the problem* $\mathrm{M}(Z, w)$, *then the function* (2.2), *where* $d\sigma(t) = P(t)d\tau(t)$, $\gamma = \rho P_0$, *is a solution of the problem* $\mathrm{NP}(Z, w)$.

PROOF. To simplify the notation, we suppose that the zeros of $P(z)$ are simple.

Let $F(z)$ be a solution of the problem $\mathrm{NP}(Z, w)$ and let (2.2) be its integral representation. If

$$\frac{t^k}{P(t)} = \sum_{z_j \in z, \operatorname{Im} z_j > 0} \left(\frac{\alpha_j^{(k)}}{t - z_j} + \frac{\overline{\alpha}_j^{(k)}}{t - \overline{z}_j} \right) + \sum_{x_j \in z, \operatorname{Im} x_j = 0} \frac{\beta_j^{(k)}}{t - x_j}, \quad k = 0, 1, \dots, N-1, \tag{2.3}$$

then

$$\int_E t^k \frac{d\sigma(t)}{P(t)} = \sum_{z_j \in z, \operatorname{Im} z_j > 0} [\alpha_j^{(k)}(F(z_j) - \gamma) + \overline{\alpha}_j^{(k)} \overline{(F(z_j) - \gamma)}] + \sum_{x_j \in z, \operatorname{Im} z_j > 0} \beta_j (F(x_j) - \gamma). \tag{2.4}$$

Since $F(z)$ is a solution of $\mathrm{NP}(Z, w)$, we may replace F by w in (2.4) and, since

$$\sum (\alpha_j^{(k)} + \overline{\alpha}_j^{(k)}) + \sum \beta_j^{(k)} = \delta_{k,N-1}/P_0,$$

we find that

$$\int_E t^k \frac{d\sigma(t)}{P(t)} = s_k - \delta_{k,N-1}\gamma/P,$$

that is, (2.1) holds with $d\tau(t) = d\sigma(t)/P(t)$ and $\rho = \gamma/P_0$.

Conversely, let (2.1) be satisfied. We may invert (2.3); for example,

$$\frac{1}{t - z_j} = \sum_{k=0}^{N-1} P_k^{(j)} \frac{t^k}{P(t)} \qquad (\operatorname{Im} z_j > 0), \tag{2.5}$$

where $P_k^{(j)}$ are the coefficients of the polynomial

$$\frac{P(t)}{t - z_j} = \sum_{k=0}^{N-1} P_k^{(j)} t^k,$$

so that $P_{N-1} = P_0$. We integrate (2.5) with respect to the measure $P(t)d\tau(t)$ and use the fact that (2.5) is the inverse of (2.3):

$$\int_E \frac{P(t)d\tau(t)}{t - z_j} = \sum_{k=0}^{N-2} P_k^{(j)} s_k + P_{N-1}^{(j)}(s_{N-1} - \rho) = w(z_j) - P_0\rho.$$

Therefore the function (2.2) with $d\sigma(t) = P(t)d\tau(t)$ and $\gamma = P_0\rho$ is a solution of the problem $\mathrm{NP}(Z, w)$. This completes the proof of the lemma.

REMARK. In the proof we used only the algebraic relations between the numbers s_k and $w(z_j)$; we made no use of the nonnegative measures $d\sigma$ and $d\tau$. Hence the conclusion of Lemma 2.1 remains valid if we drop the nonnegativity of the measures.

§3. Material from the theory of moments

In view of the discovery of the connection between the measures that yield solutions of the $\mathrm{NP}(Z, w)$ and $\mathrm{M}(Z, w)$ problems, it makes sense to apply the rather thoroughly developed machinery of the theory of moments. We present the material that is needed for what follows, referring to [3] for details.

If a sequence $\{s_k\}_0^{N-1}$ has the representation

$$s_k = \int_0^\infty t^k\, d\tau(t) + \rho\delta_{N-1,k} \qquad (k = 0, 1, \dots, N-1;\ d\tau(t) \geq 0,\ \rho \geq 0), \quad (3.1)$$

where $d\tau$ has infinite support, then there is an infinite set of measures that satisfy (3.1). These include two simplest measures, known as *principal* measures. The structure of the principal measures can be described as follows. They both have finite support. If a finite mass of $d\tau$ is concentrated at a point other than 0, the point is assigned index 2; the point 0 is assigned index 1. The number ρ in (3.1) is treated as "the mass at ∞", and when $\rho > 0$ the point at ∞ is assigned index 1. For the measure $d\tau$ determined by the representation (3.1) of the sequence $\{s_k\}_0^{N-1}$, we say that it *determines a principal representation* if the sum of the indices of the points where the mass of this measure is concentrated (including ∞) equals N. We call a principal representation a *first principal representation* if there is mass at the point 0, and a *second principal representation* if there is no mass at 0.(2) The first and second principal representations are uniquely determined by $\{s_k\}_0^{N-1}$. Between any two successive points where mass of a principal representation is concentrated there is at least one point of increase of any measure $d\tau$ that yields a representation (3.1). The finite points of concentration of the first or second principal representation are zeros of the polynomial

$$q_0(z) = c_0 z \det \|z^k s_{k+1} s_{k+2} \cdots s_{k+\mu}\|_{k=0}^{\mu} \qquad (\mu = [(N-1)/2]), \qquad (3.2)$$

or

$$q_1(z) = c_1 \det \|z^k s_k s_{k+1} \cdots s_{k+\nu-1}\|_{k=0}^{\nu} \qquad (\nu = [N/2]), \qquad (3.3)$$

respectively.

To each polynomial $q(z)$ of degree $\leq N-1$ we assign an *associated* polynomial(3) $p(z)$ according to the rule: in $(q(t) - q(z))/(t - z)$, replace each power t^k by $(-s_k)$. In other words,

$$p(z) = -\int_0^\infty \frac{q(t) - q(z)}{t - z}\, d\tau(t),$$

where $d\tau(t)$ is any solution of the moment problem (3.1). It is easily seen that if the measure $d\tau(t)$ contains only point masses, at the points ξ_j, and all these points are zeros of $q(z)$, then

$$\frac{p(z)}{q(z)} = \sum_j \frac{\rho_j}{\xi_j - z}.$$

(2)In [3] the principal representations are called upper or lower according as there is or is not mass at ∞.

(3)The "conjugate" polynomials introduced in [3] have the sign opposite to that of the associated polynomials.

The polynomials associated with $q_0(z)$ and $q_1(z)$ will be denoted by $p_0(z)$ and $p_1(z)$. We normalize these four polynomials by choosing the constants c_0 and c_1 in (3.2) and (3.3) so that $p_0(0) = 1$ and $q_1(0) = 1$. Then we have the identity

$$p_0(z)q_1(z) - p_1(z)q_0(z) = 1. \tag{3.4}$$

The formula

$$\frac{p_0(z)\omega(z) + p_1(z)}{q_0(z)\omega(z) + q_1(z)} = \int_0^\infty \frac{d\tau(t)}{t - z} \tag{3.5}$$

establishes a one-to-one correspondence between measures $d\tau(t)$ that have the given moments and functions $\omega(z) \in S$ (including $\omega(z) = \infty$).

§4. First solution of Problem A_0

For a polynomial $P(z)$, positive on $E = [0, \infty)\backslash \bigcup_1^g (a_j, b_j)$, we may suppose without loss of generality that $P(0) = 1$. The representation

$$P(z) = A^2(z) + T(z)B^2(z),$$

where $T(z) = T_E(z) = za(z) \cdot b(z)$, is said to be *special* if $A(z)$ and $B(z)$ are a special pair of polynomials for which $A(0) = 1$ and the number N [see §1.1, b) for the definition of a special pair] equals the degree of $P(z)$.

Problem A_0 is to find necessary and sufficient conditions for a polynomial $P(z)$, positive on E, to have a special representation.

It is easily seen that the function $-a(z)/(zb(z))$ belongs to class $S_+(E)$. So does the function

$$v(z) = \sqrt{-a(z)/(zb(z))}; \tag{4.1}$$

where we take the branch on $\mathbb{C}\backslash E$ that is positive on $(-\infty, 0)$.

By using the inversion formula (1.1), it is easy to see that

$$v(z) = \frac{1}{\pi} \int_E \frac{\sqrt{a(t)/(tb(t))}}{t - z}\, dt. \tag{4.2}$$

In what follows, $\{s_k\}_0^{N-1}$ will denote the moment sequence of the problem $\mathrm{M}(Z, v)$. This sequence admits the representation

$$s_k = \frac{1}{\pi} \int_E t^k \frac{\sqrt{a(t)/(tb(t))}}{P(t)}\, dt, \qquad k = 0, 1, \dots, N-1. \tag{4.3}$$

THEOREM 4.1. *A necessary and sufficient condition for the polynomial* $P(z)$, *positive on* E, *to admit the special representation*

$$P(z) = A^2(z) + T(z)B^2(z), \tag{4.4}$$

is that the conditions

$$q_0(b_k) = 0, \qquad k = 1, 2, \dots, g,$$

are satisfied. In this case the representation (4.4) *is unique and*

$$A(z) = q_1(z), \qquad B(z) = \frac{q_0(z)}{zb(z)}.$$

PROOF. a) Let (4.4) hold. Then

$$P(z) = z^2 b^2(z) B^2(z) \left(\frac{A(z)}{zb(z)B(z)} - v(z) \right) \left(\frac{A(z)}{zb(z)B(z)} + v(z) \right).$$

Since $A(z)/(zb(z)B(z)) \in S_+(E)$, the preceding factors do not have zeros in $\mathbb{C} \setminus E$. Hence it follows that $A(z)/(zb(z)B(z))$ is a solution of the problem $\mathrm{NP}(Z, v)$. If we expand this quotient in partial fractions, we obtain

$$\frac{A(z)}{zb(z)B(z)} = \gamma + \sum_{j=1}^{[(N+1)/2]} \frac{\mu_j}{\xi_j - z}, \tag{4.5}$$

where N is the degree of $P(z)$, $\mu_j > 0$, the ξ_j are the zeros of $zb(z)B(z)$, $\gamma > 0$ for $N = 2n$, and $\gamma = 0$ for $N = 2n + 1$. One of the ξ_j is 0, and the sum of the indices of the points ξ_j and ∞ for $\gamma > 0$ is equal to N. These points, by Lemma 2.1, are points of increase of the solution of problem $\mathrm{M}(Z, v)$. Therefore the zeros of $zb(z)B(z)$ are points that carry mass of the first principal representation of the moments $\{s_k\}_0^{N-1}$ and consequently coincide with the zeros of $q_0(z)$. Hence it follows that $q_0(b_k) = 0$, $k = 1, 2, \ldots, g$. Moreover, in this case the polynomials $zb(z)B(z)$ and $q_0(z)$ can differ only by a constant factor; presently we shall prove that they are the same.

b) Conversely, let $q_0(b_k) = 0$, $k = 1, 2, \ldots, g$. Let ξ_j be the zeros of $q_0(z)$. Since these include the right-hand endpoints of the intervals (a_k, b_k), none of the points ξ_j can lie in these intervals or at their left-hand endpoints. In fact, if we had $\xi_j \in [a_k, b_k)$, then (a_k, b_k) would contain points of increase of every solution of the moment problem $\mathrm{M}(Z, v)$. However, the sequence $\{s_k\}_0^{N-1}$ admits the representation (4.3), in which the solution has no points of increase in any interval (a_k, b_k).

We set $B(z) = q_0(z)/(zb(z))$ and

$$A(z) = zb(z)B(z) \left(\rho P_0 + \sum_{j=1}^{[(N+1)/2]} \frac{\rho_j P(\xi_j)}{\xi_j - z} \right), \tag{4.6}$$

where ρ_j and ρ are the masses in the first principal representation. It is clear that $A(z)$ and $B(z)$ are polynomials of the required degrees. Since all $\xi_j \in E$, the quotient $A(z)/(zb(z)B(z)) \in S(E)$; moreover, by Lemma 2.1, this quotient is a solution of the problem $\mathrm{NP}(Z, v)$. Let us show that $A(z)/(zb(z)B(z)) \in S_+(E)$. It is enough to show that

$$(b(z)/a(z))A(z)/(zb(z)B(z)) = A(z)/(za(z)B(z)) \in S.$$

Let β_j be the zeros of $za(z)B(z)$. They all belong to $E \subset [0, \infty)$; hence it remains only to prove that all $\widetilde{\mu}_j > 0$ in the decomposition

$$\frac{A(z)}{za(z)B(z)} = \rho P_0 + \sum_{j=1}^{[(N+1)/2]} \frac{\widetilde{\mu}_j}{\beta_j - z}. \tag{4.7}$$

Set $\widetilde{v}(z) = -1/(zv(z))$ ($\in S(E)$) and observe that Z is the set of zeros of the difference

$$\frac{A(z)}{za(z)B(z)} - \widetilde{v}(z) = \frac{b(z)}{a(z)}\left(\frac{A(z)}{zb(z)B(z)} - v(z)\right).$$

Therefore, if we set $\widetilde{\rho}_j = \widetilde{u}_j/P(\beta_j)$ we have, using the remark on Lemma 2.1, the representation

$$\widetilde{s}_k = \sum_{j=1}^{[(N+1)/2]} \widetilde{\rho}_j \beta_j^k + \delta_{k,N-1}\rho, \qquad k = 0, 1, \dots, N-1,$$

for $\{\widetilde{s}_k\}_0^{N-1}$.

Since the sum of the indices of the points β_j and ∞ (if $\rho > 0$) is N, we may construct a polynomial $Q_l(t) = \sum_{k=0}^{N-1} Q_{lk}t^k$ which is nonnegative on $[0, \infty)$ and has the property that $Q_l(\beta_j) = 0$ for $j \neq l$, $Q_l(\beta_l) = 1$, and $Q_{l,N-1} = 0$ if $\rho > 0$. For this polynomial

$$\sum_{k=0}^{N-1} Q_{lk}\widetilde{s}_k = \sum_{j=1}^{[(N+1)/2]} \widetilde{\rho}_j Q_l(\beta_j) + \delta_{k,N-1}\rho Q_{l,N-1} = \widetilde{\rho}_l.$$

On the other hand, since there is a representation

$$\widetilde{s}_k = \int_0^\infty t^k\, d\widetilde{\tau}(t), \qquad k = 0, 1, \dots, N-1,$$

where the measure $d\widetilde{\tau}(t) \geq 0$ and has infinite support, we have

$$\widetilde{\rho}_l = \sum_{k=0}^{N-1} Q_{lk}\widetilde{s}_k = \int_0^\infty \sum_{k=0}^{N-1} Q_{lk}t^k\, d\widetilde{\tau}(t) = \int_0^\infty Q_l(t)\, d\widetilde{\tau}(t) > 0.$$

Consequently $\widetilde{\mu}_l = \widetilde{\rho}_l P(\rho_l) > 0$; that is, $A(z)/(zb(z)B(z)) \in S_+(E)$.

Let us show that $A(0) = 1$. For this purpose we use (4.6), where $zb(z)B(z) = q_0(z)$, and the relation

$$\frac{p_0(z)}{q_0(z)} = \sum_{j=1}^{[(N+1)/2]} \frac{\rho_j}{\xi_j - z} \qquad (0 = \xi_1 < \xi_2 < \cdots). \tag{4.8}$$

Since $\xi_1 = 0$, we find from (4.6) that $-A(0)/q'(0) = \rho_1 P(0) = \rho_1$, and from (4.8), that $\rho_1 = -p(0)/q'(0) = -1/q_0'(0)$, whence $A(0) = 1$.

c) Thus, under the conditions $q_0(b_k) = 0$, $k = 1, 2, \dots, g$, we have constructed polynomials $A(z)$ and $B(z)$, of respective degrees $[N/2]$ and $[(N-1)/2]$, such that $A(0) = 1$ and $A(z)/(zb(z)B(z)) \in S_+(E)$. This fraction is a solution of problem $\mathrm{NP}(Z, v)$. Consequently the set of zeros of the function

$$\left(\frac{A(z)}{zb(z)B(z)} - v(z)\right)\left(\frac{A(z)}{zb(z)B(z)} + v(z)\right) = \frac{A^2(z) + T(z)B^2(z)}{z^2b^2(z)B^2(z)}$$

and consequently of the polynomial

$$A^2(z) + T(z)B^2(z) \tag{4.9}$$

contains the zeros of $P(z)$. Since (4.9) is of degree N and its value at $z=0$ is $A^2(0)=1=P(0)$, it coincides with $P(z)$.

d) It remains to establish the uniqueness of the representation (4.4). It has been shown that in the presence of this representation the polynomial $zb(z)B(z)$ differs from $q_0(z)$ at most by a constant factor. Their identity will be established if it turns out that the polynomial $C(z)$ associated with the polynomial $zb(z)B(z)$ satisfies the equation $C(0)=1$.

From the decomposition (4.5) we obtain that $\mu_1 = -A(0)/(b(0)B(0)) = -1/(b(0)B(0))$. On the other hand, by Lemma 2.1 and the definition of the associated polynomial,

$$\frac{C(z)}{zb(z)B(z)} = \sum_{j=1}^{[(N+1)/2]} \frac{\rho_j}{\xi_j - z},$$

where $\rho_j = \mu_j/P(\xi_j)$. Hence $\mu_1 = \rho_1 P(0) = \rho_1 = -C(0)/(b(0)B(0))$, that is, $C(0)=1$.

Finally let us prove that $A(z)=q_1(z)$. Since $F(z)=A(z)/(zb(z)B(z)) \in S_+(E)$, we have $G(z)=(zb(z)/a(z))F(z) \in \mathscr{R}$ and, as is easily seen, $H(z) = -a(z)B(z)/A(z) = -1/G(z) \in S_+(E)$. Since

$$\left(-\frac{a(z)B(z)}{A(z)}\right) - v(z) = \frac{a(z)B(z)}{A(z)v(z)}\left(\frac{A(z)}{zb(z)B(z)} - v(z)\right),$$

the function $-a(z)B(z)/A(z)$ is also a solution of the problem $\mathrm{NP}(Z, v)$. Therefore the set of zeros of $A(z)$ supports a measure which is a solution of the problem $\mathrm{M}(Z, v)$. It remains only to observe that the sum of the indices of these zeros and the point ∞ (if the degrees of the polynomials $a(z)B(z)$ and $A(z)$ are the same) is N, that is, the zeros of $A(z)$ are the finite points where mass of the second principal representation of the moments $\{s_k\}_0^{N-1}$ is concentrated. Therefore $A(z)=cq_1(z)$, and since $A(0)=1$ and $q_1(0)=1$, we have $A(z)=q_1(z)$. This completes the proof of the theorem.

§5. The second solution of Problem A_0

1. In this section we shall obtain a second form of the criterion for the existence of a special representation. At the same time we find a method for determining the set of zeros of the polynomial $A(z)$ in the intervals $(0, a_k)$ by means of certain integrals. Moreover, we obtain an expression that lets us present $A(z)$ and $B(z)$ in terms of the zeros of $P(z)$.

As a preliminary we observe that the determinant of order g with elements $\alpha_{jk} = \int_{a_k}^{b_k} x^{j-1}\,dx/\sqrt{|T(x)|}$, $j, k = 1, 2, \dots, g$, is not zero. To verify this, it is sufficient to integrate the (positive) Vandermonde determinant $\det \|x_k^{j-1}\|$ $(j, k = 1, 2, \dots, g;\ x_k \in [a_k, b_k])$ g times over the intervals $[a_k, b_k]$ with respect to the measure $dx_k/\sqrt{|T(x_k)|}$.

We cut $\mathbb{C}$ along E; the indices "u" and "l" will denote the upper and lower sides of the cut. By the remarks made above, the numbers n_k in the relations

$$\frac{1}{2\pi}\int_{E_u}\frac{x^j\ln P(x)}{\sqrt{T(x)}}\,dx=\sum_{k=1}^{g}n_k\int_{a_k}^{b_k}\frac{x^j\,dx}{\sqrt{|T(x)|}}\quad(j=0,1,\dots,g-1)\tag{5.1}$$

are uniquely determined.

THEOREM 5.1. *A necessary and sufficient condition for a polynomial $P(z)$, positive on E, to have the special representation*

$$P(z)=A^2(z)+T(z)B^2(z),\tag{5.2}$$

is that the numbers n_k defined by (5.1) *are integers.*

Under these hypotheses the number $\nu_k=(-1)^k n_k\geq 0$ and equals the number of zeros of $A(z)$ in the interval $(0,a_k)$, $k=1,2,\dots,g$.

PROOF. *Necessity.* Let the special pair $A(z)$, $B(z)$ have the representation (5.2). Then

$$\begin{aligned}\frac{1}{2\pi}\int_{E_u}\frac{x^j\ln P(x)}{\sqrt{T(x)}}\,dx&=\frac{1}{\pi}\operatorname{Re}\int_{E_u}\frac{x^j\ln(A(x)-i\sqrt{T(x)}B(x))}{\sqrt{T(x)}}\,dx\\&=\frac{1}{\pi}\left(\operatorname{Re}\int_{E_u}\frac{x^j\ln A(x)}{\sqrt{T(x)}}\,dx+\operatorname{Re}\int_{E_u}\frac{x^j\ln(1-i\sqrt{T(x)}B(x)/A(x))}{\sqrt{T(x)}}\,dx\right).\end{aligned}$$

Since $\operatorname{Im}(\sqrt{T(z)}B(z)/A(z))>0$ for $z\in\mathbb{C}\setminus E$, the function $w=1-i\sqrt{T(z)}B(z)/A(z)$ maps $\mathbb{C}\setminus E$ on the half plane $\operatorname{Re}w>1$. Choosing a holomorphic branch of $\ln w=\ln|w|+i\arg w$ in this half plane (for example, taking $|\arg w|<\pi/2$), we have $\ln(1-i\sqrt{T(z)}B(z)/A(z))$ holomorphic in the multiply-connected domain $\mathbb{C}\setminus E$. Hence, in particular, it follows that its values on $(a_k,b_k)_u$ and $(a_k,b_k)_l$ are equal, $k=1,2,\dots,g$.

We take the limit as $r\to\infty$ in the equation

$$\int_{C_r}\frac{z^j\ln(1-i\sqrt{T(z)}B(z)/A(z))}{\sqrt{T(z)}}\,dz=0,$$

where C_r is the contour made up of the interval $(0,r)_u$, the circle of radius r with center at 0, and the interval $(r,0)_l$. The integral over the circle tends to 0. Therefore, since $\sqrt{T(z)}|_{E_u}=-\sqrt{T(z)}|_{E_l}$ and $\sqrt{T(z)}|_{(a_k,b_k)_u}=\sqrt{T(z)}|_{(a_k,b_k)_l}$, $k=1,2,\dots,g$, we obtain in the limit

$$\int_{E_u}\frac{x^j\ln(1-i\sqrt{T(x)}B(x)/A(x))}{\sqrt{T(x)}}\,dx=0.$$

Therefore

$$\frac{1}{2\pi}\int_{E_u}\frac{x^j\ln P(x)\,dx}{\sqrt{T(x)}}=\frac{1}{\pi}\int_{E_u}\frac{x^j\ln|A(x)|}{\sqrt{T(x)}}\,dx.$$

To evaluate the last integral we observe that in the simply-connected domain bounded by C_r the function $z^j \ln A(z)/\sqrt{T(z)}$ is holomorphic (all the zeros of $A(z)$ lie in E). Therefore

$$\int_{C_r} \frac{z^j \ln A(z)}{\sqrt{T(z)}}\, dz = 0 \qquad (j = 0, 1, \dots, g-1).$$

When $r \to \infty$ we obtain in the limit

$$\left(\int_{(0,\infty)_u} - \int_{(0,\infty)_l}\right) \frac{x^j \ln A(x)}{\sqrt{T(x)}}\, dx = 0,$$

so that

$$\operatorname{Re}\left[\left(\int_{E_u} - \int_{E_l}\right) \frac{x^j \ln A(x)}{\sqrt{T(x)}}\, dx + \sum_{k=1}^{g}\left(\int_{(a_k,b_k)_u} - \int_{(a_k,b_k)_l}\right) \frac{x^j \ln A(x)}{\sqrt{T(x)}}\, dx = 0. \tag{5.3}$$

It is easily seen that $\ln A(z)|_{(a_k,b_k)_u} = \ln|A(z)| - \nu_k \pi i$, where ν_k is the number of zeros of $A(z)$ in $(0, a_k)$. After a circuit around the circumference C_r the argument of $A(z)$ is decreased by $2\pi n i$ (n is the degree of $A(z)$) and consequently

$$\ln A(z)|_{(a_k,b_k)_l} = (\ln|A(z)| - n\pi i) + 2n\pi i - (n - \nu_k)\pi i.$$

Moreover,

$$\sqrt{T(z)}|_{(a_k,b_k)_l} = \sqrt{T(z)}|_{(a_k,b_k)_u}.$$

Taking account of this, we find from (5.2) that

$$\frac{1}{\pi}\int_{E_u} \frac{x^j \ln|A(x)|}{\sqrt{T(x)}}\, dx = \sum_{k=1}^{g} \nu_k i \int_{a_k}^{b_k} \frac{x^j\, dx}{\sqrt{T(x)}}, \qquad j = 0, 1, \dots, g-1.$$

To obtain (5.1) it is enough to notice that

$$\sqrt{T(x)}|_{(a_k,b_k)} = (-1)^k i \sqrt{|T(x)|}.$$

Sufficiency. Let the numbers n_k determined by (5.1) be integral.

a) We construct the Riemann surface $\mathcal{R}$ of $w = \sqrt{T(z)}$. For this purpose we superimpose two sheets $\mathbb{C}_E^+$ and $\mathbb{C}_E^-$, cut along E. We denote by E_u^+ and E_l^+ the upper and lower edges of the cuts on $\mathbb{C}_E^+$, and by E_u^- and E_l^- the upper and lower edges on $\mathbb{C}_E^-$. Joining E_u^+ to E_l^- and E_u^- to E_l^+, we obtain $\mathcal{R}$. The branch points lie over the points $0, a_1, b_1, \dots, a_g, b_g, \infty$. As we know, the genus of the surface $\mathcal{R}$ is g. The integrals

$$\mathcal{J}^j(\gamma) = \int_\gamma \frac{z^j}{\sqrt{T(z)}}\, dz, \qquad j = 0, 1, \dots, g-1, \tag{5.4}$$

where γ is a path on $\mathcal{R}$, form a complete set of linearly independent Abelian integrals of first order on $\mathcal{R}$, that is, every Abelian integral of the first order on $\mathcal{R}$ is a linear combination of the integrals (5.4).

We denote by A_k a simple contour on $\mathbb{C}_E^+$ surrounding the interval (b_{k-1}, a_k), $k = 1, 2, \dots, g$ ($b_0 = 0$), and by B_k a simple contour surrounding

(a_k, b_k), $k = 1, 2, \ldots, g$, that passes from the lower half plane of $\mathbb{C}_E^+$ to the upper half plane of $\mathbb{C}_E^-$. These contours can be regarded as a canonical system of cross-cuts of $\mathcal{R}$. When the path γ coincides with one of these cuts A_k or B_k, the corresponding integral $\mathcal{J}^j(\gamma)$ is of period ω_k^j or $\widehat{\omega}_k^j$. Since

$$\sqrt{T(z)}|_{E_u^+} = \sqrt{T(z)}|_{E_l^-} \quad \text{and} \quad \sqrt{T(z)}|_{(a_k, b_k)^+} = -\sqrt{T(z)}|_{(a_k, b_k)^-},$$

we have

$$\omega_k^j = -2\int_{(b_{k-1}, a_k)^+} \frac{x^j\,dx}{\sqrt{T(x)}}, \quad \widehat{\omega}_k^j = 2\int_{(a_k, b_k)^+} \frac{x^j\,dx}{\sqrt{T(x)}} = (-1)^k 2i\int_{a_k}^{b_k} \frac{x^j\,dx}{\sqrt{|T(x)|}}.$$

Every closed path γ on the Riemann surface $\mathcal{R}$ is homologous to a combination $\sum_1^g (n_k' A_k + n_k'' B_k)$ and, correspondingly, $\mathcal{J}^j(\gamma) = \Omega^j$, where

$$\Omega^j = \sum_{k=1}^{g} (n_k' \omega_k^j + n_k'' \widehat{\omega}_k^j), \qquad j = 0, 1, \ldots, g-1.$$

The numbers Ω^j (linear combinations with integral coefficients) form a full period module of the Abelian integral.

Condition (5.1) shows that the vector $\vec{L}(P) = (L^0(P), L^1(P), \ldots, L^{g-1}(P))$, where

$$L^j(P) = \frac{1}{\pi i}\int_{E_u^+} \frac{x^j \ln P(x)\,dx}{\sqrt{T(x)}}$$

coincides with the period vector

$$\vec{\Omega} = (\Omega^0, \Omega^1, \ldots, \Omega^{g-1}) = \sum_{k=1}^{g} \nu_k (\widehat{\omega}_k^0, \widehat{\omega}_k^1, \ldots, \widehat{\omega}_k^{g-1}),$$

where $\nu_k = (-1)^k n_k$.

b) Let $\{\alpha_k^+\}_1^N$ denote the points on $\mathbb{C}_E^+$ that project to the zeros $\{\alpha_k\}_1^N$ of $P(z)$. We join these points to the point at infinity by rays $\{\gamma_k\}_1^N$, not intersecting E. It is easily seen that we have the equation

$$\vec{L}(P) = \sum_{k=1}^{N} \vec{\mathcal{J}}(\gamma_k),$$

where

$$\vec{\mathcal{J}}(\gamma_k) = (\mathcal{J}^0(\gamma_k), \mathcal{J}^1(\gamma_k), \ldots, \mathcal{J}^{g-1}(\gamma_k)). \tag{5.5}$$

In fact, let us make cuts along γ_k and denote by D_r the intersection of the resulting domain $D \subset \mathbb{C}_E^+$ with the disk $|z| < r$. Integrating $z^j \ln P(z)/\sqrt{T(z)}$ over the whole correctly oriented boundary ∂D_r, and passing to the limit as $r \to \infty$, we obtain (5.5).

Thus for a given set of points $\{\alpha_k^+\}_1^N$ and rays $\{\gamma_k\}_1^n$ we have $\sum_1^n \vec{\mathcal{J}}(\gamma_k) = \vec{\Omega}$. Consequently, by Abel's theorem(4), there is on $\mathcal{R}$ a meromorphic func-

(4)We use the formulation of Abel's theorem in [4]. A better known formulation is that in which one requires the existence of simple contours $\widetilde{\gamma}_k$ for which $\sum_1^n \vec{\mathcal{J}}(\widetilde{\gamma}_k) = 0$. It is clear that these requirements are equivalent.

tion $f(z)$ with zeros at the points $\{\alpha_k^+\}_1^N$ and a pole of order N at infinity. The function $f(z)$ is determined up to a multiplicative constant, since the quotient of two such functions has no singular points on $\mathcal{R}$.

We now show that, with a suitable choice of the constant multiplier, $f(z)$ takes real values on the set F^+ of intervals of the real axis complementary to E on $\mathbb{C}_E^+$.

In fact, let $\alpha \in F^+$ and in a neighborhood of this point let

$$f(z) = \sum_{n=0}^{\infty} c_n (z - \alpha)^n.$$

The function

$$\overline{f}(z) = \sum_{n=0}^{\infty} \overline{c}_n (z - \alpha)^n$$

can be continued along a path symmetric with respect to the real axis to the path along which $f(z)$ is continued on $\mathcal{R}$, and at the symmetric point z_* (the end of the symmetric path) it takes the conjugate value: $\overline{f}(z_*) = \overline{f(z)}$. Therefore $\overline{f}(z)$ is meromorphic on $\mathcal{R}$, is zero at the points $\{\overline{\alpha}_k^+\}_1^N$, and has a pole of order n at infinity. Since the zeros of $P(z)$ are symmetrically distributed with respect to the real axis, the set $\{\overline{\alpha}_k^+\}_1^N$ coincides with the set $\{\alpha_k^+\}_1^N$. Therefore if we choose $f(z)$ so that the number c_0 is real, we obtain, by uniqueness, $\overline{f}(z) = f(z)$ $(z \in \mathcal{R})$. Hence it follows that $f(z)$ is real on F^+. It is also evident that $f(z)$ will take real values on F^- in $\mathbb{C}_E^-$. On E_u^+ and E_l^+, $f(z)$ takes complex conjugate values: $f(z_u^+) = \overline{f(z_l^+)}$. By Abel's theorem, the set of zeros of $f(z)$ exhausts the set $\{\alpha_k^+\}_1^N$, since its continuation across E on $\mathbb{C}_E^-$ cannot have any zeros there. Finally we notice that by the way the sheets are interconnected we will have

$$f(z_u^-) = \overline{f(z_u^+)}; \qquad f(z_l^-) = \overline{f(z_l^-)}. \tag{5.6}$$

c) Now we interchange the roles of $\mathbb{C}_E^+$ and $\mathbb{C}_E^-$ and construct, in a similar way, a meromorphic function $f_*(z)$ with zeros at $\{\alpha_k^-\}_1^N$ and a pole of order N at infinity. It has no zeros on $\mathbb{C}_E^+$, is real on the intervals F^+ and F^- of the real axis, and takes, at the points of E, values which (as is clear from (5.6)) are the conjugates of the values of $f(z)$. Moreover, on a circuit around any branch point of $\mathcal{R}$ the functions $f(z)$ and $f_*(z)$ are interchanged. Evidently the function $Q(z) = f(z)f_*(z)$ takes equal values at the points z^+ and z^- of $\mathcal{R}$; that is, it is holomorphic on $\mathbb{C}$, vanishes at $\{\alpha_k\}_1^N$, and has a pole of order N at infinity. Therefore, if we choose the constant real multiplier of $f(z)$ in the corresponding way, we will have

$$P(z) = f(z)f_*(z). \tag{5.7}$$

The function $A(z) = \frac{1}{2}(f(z)+f_*(z))$ also does not change on a circuit around any branch point of $\mathcal{R}$, and therefore is a real polynomial of degree $\leq N/2$.

It is easily verified that the function

$$B(z) = \frac{1}{2i}\frac{f(z) - f_*(z)}{\sqrt{T(z)}}$$

takes equal values at z^+ and z^-, is real on the real axis, and has removable singularities at the points 0, a_k, and b_k ($k = 1, 2, \dots, g$). Therefore $B(z)$ is a real polynomial, of degree not exceeding $(N-1)/2 - g$.

Thus

$$f(z) = A(z) + i\sqrt{T(z)}B(z),$$
$$f_*(z) = A(z) - i\sqrt{T(z)}B(z).$$

Formula (5.7) can be rewritten in the form

$$P(z) = A^2(z) + T(z)B^2(z).$$

d) It remains to verify that $A(z)$ and $B(z)$ form a special pair. For this purpose we consider the function

$$\varphi(z) = \frac{f(z)}{f_*(z)}$$

on $\mathbb{C}_E^+$; this is holomorphic on the cut sheet $\mathbb{C}_E^+$ and has a removable singularity at infinity; also its modulus has boundary values 1 at all points of E. Since $f(z)$ has zeros only in $\operatorname{Int}\mathbb{C}_E^+$, we have $|\varphi(z)| < 1$ for $z \in \mathbb{C}_E^+ = \mathbb{C}\backslash E$. By Lemma 1.1, this means that $A(z)$ and $B(z)$ form a special pair. This completes the proof of the theorem.

COROLLARY. *The positive polynomials $P(z)$ on E that admit the representation* (5.2) *form a semigroup $\mathfrak{P}$, that is, if P_1, $P_2 \in \mathfrak{P}$, then $P_1P_2 \in \mathfrak{P}$. Moreover, if $P_1 \in \mathfrak{P}$ and $P_1P_2 \in \mathfrak{P}$, then $P_2 \in \mathfrak{P}$.*

This follows immediately from Theorem 5.1 and the equation $\vec{L}(P_1P_2) = \vec{L}(P_1) + \vec{L}(P_2)$.

2. In §4 we obtained a procedure for constructing $A(z)$ and $B(z)$. In the present subsection we give a different method for the construction, and obtain some new formulas for $A(z)$ and $B(z)$, which may be of independent interest.

We first define two functions:

$$\varphi(z) = \frac{f'(z)}{f(z)} - \sum_{j=1}^{N} \frac{\sqrt{T(z)} + \sqrt{T(\alpha_j^+)}}{2(z-\alpha_j)\sqrt{T(z)}} \tag{5.8}$$

and

$$\varphi_*(z) = \frac{f_*'(z)}{f_*(z)} - \sum_{j=1}^{N} \frac{\sqrt{T(z)} + \sqrt{T(\alpha_j^-)}}{2(z-\alpha_j)\sqrt{T(z)}}, \tag{5.9}$$

where each term on the right is repeated according to the multiplicity of the root α_j of $P(z)$.

We normalize $\sqrt{T(z)}$ on $\mathcal{R}$ by the condition $T(x_u^+) > 0$ in $(0, a_1)$. Since $\sqrt{T(\alpha^-)} = -\sqrt{T(\alpha^+)}$, the functions (5.8) and (5.9) have no singular points in the finite part of $\mathcal{R}$, except perhaps for simple poles at the zeros of $T(z)$. In a neighborhood of infinity the functions $\varphi(z)$ and $\varphi_*(z)$ satisfy the condition

$$\lim_{z\to\infty} z\varphi(z) = \lim_{z\to\infty} z\varphi_*(z) = 0. \tag{5.10}$$

In fact, as is easily seen,

$$\lim_{z\to\infty} \frac{zf'(z)}{f(z)} = \lim_{z\to\infty} \frac{zf_*'(z)}{f_*(z)};$$

on the other hand,

$$\lim_{z\to\infty} z\left(\frac{f'(z)}{f(z)} + \frac{f_*'(z)}{f_*(z)}\right) = \lim_{z\to\infty} z\frac{P'(z)}{P(z)} = N$$

and

$$\lim_{z\to\infty} z\sum_{j=1}^{N} \frac{\sqrt{T(z)} + \sqrt{T(\alpha_j^{\pm})}}{2(z-\alpha_j)\sqrt{T(z)}} = \frac{N}{2},$$

from which (5.10) follows. Therefore, both functions are holomorphic in a neighborhood of infinity on $\mathcal{R}$, and can be expanded in this neighborhood in terms of the local parameter $\zeta = \sqrt{z}$ as follows:

$$\varphi(\zeta) = \frac{c_3}{\zeta^3} + \frac{c_4}{\zeta^4} + \cdots; \qquad \varphi_*(\zeta) = \frac{c_3^*}{\zeta^3} + \frac{c_4^*}{\zeta^4} + \cdots.$$

From the equation

$$\varphi(z) + \varphi_*(z) = \frac{P'(z)}{P(z)} - \sum_{j=1}^{N} \frac{1}{z-\alpha_j} = 0$$

it follows that $c_\nu = -c_\nu^*$ $(\nu = 3, 4, \dots)$. Now we consider the difference $\varphi(z) - \varphi_*(z)$, which can be represented in the form

$$i\frac{2ATB' - 2A'TB - ABT'}{P(z)\sqrt{T(z)}} - \sum_{j=1}^{N} \frac{\sqrt{T(\alpha_j^+)}}{(z-\alpha_j)\sqrt{T(z)}} = \frac{2c_3}{\zeta^3} + \frac{2c_4}{\zeta^4} + \cdots.$$

The left-hand side of this equation is an odd function of ζ, and hence all $c_{2p} = 0$. The second term on the left-hand side decreases at infinity with order $O(|\zeta|^{-2g-3})$; hence all the coefficients $c_3, c_5, \dots, c_{2g+1}$ are pure imaginaries. Setting $c_\nu = ie_\nu$, we obtain

$$\frac{f'(z)}{f(z)} = \sum_{j=1}^{N} \frac{\sqrt{T(z)} + \sqrt{T(\alpha_j^+)}}{(z-\alpha_j)\sqrt{T(z)}} + \frac{ie_3}{\zeta^3} + \frac{ie_5}{\zeta^5} + \cdots + \frac{ie_{2g+1}}{\zeta^{2g+1}} + O(|\zeta|^{-2g-3}). \tag{5.11}$$

We now observe that the derivatives of the integrals of first order have the asymptotic expansions

$$\frac{z^{g-1}}{\sqrt{T(z)}} = \frac{1}{\zeta^3} + \frac{f_5}{\zeta^5} + \cdots; \quad \frac{z^{g-2}}{\sqrt{T(z)}} = \frac{1}{\zeta^5} + \cdots; \quad \cdots; \quad \frac{1}{\sqrt{T(z)}} = \frac{1}{\zeta^{2q+1}} + \cdots.$$

It is evident that, using (5.11), we can successively select the real coefficients d_ν so that we have the asymptotic formula

$$\frac{f'(z)}{f(z)} - \sum_{j=1}^{N} \frac{\sqrt{T(z)} + \sqrt{T(\alpha_j^+)}}{2(z-\alpha_j)\sqrt{T(z)}} - i\frac{d_0 + d_1 z + \cdots + d_{g-1}z^{g-1}}{\sqrt{T(z)}} = O(|\zeta|^{-2g-3}).$$

The function on the left-hand side is meromorphic on $\mathcal{R}$, has $2g+1$ simple poles at the zeros of $T(z)$, and has a zero of order not less than $2g+3$ at infinity. Hence it follows that it is identically zero.

Thus

$$\frac{f'(z)}{f(z)} = \sum_{j=1}^{N} \frac{\sqrt{T(z)} + \sqrt{T(\alpha_j^+)}}{2(z-\alpha_j)\sqrt{T(z)}} + i\frac{d_0 + d_1 z + \cdots + d_{g-1}z^{g-1}}{\sqrt{T(z)}}. \tag{5.12}$$

There is a similar representation for

$$\frac{f_*'(z)}{f_*(z)} = \sum_{j=1}^{N} \frac{\sqrt{T(z)} + \sqrt{T(\alpha_j^+)}}{2(z-\alpha_j)\sqrt{T(z)}} + i\frac{d_0 + d_1 z + \cdots + d_{g-1}z^{g-1}}{\sqrt{T(z)}}. \tag{5.12$'$}$$

We have therefore shown that the logarithmic derivative of the required function $f(z)$ is represented by (5.12), and the problem reduces to determining the real coefficients $d_0, d_1, \ldots, d_{g-1}$. We can find these by using the fact that $\ln f(z)$ has no periods with real part different from zero. We first observe that when $z \in (-\infty, \infty)\backslash E$ the function $\sqrt{T(z)}$ is pure imaginary and that therefore the second term on the right of (5.12) is a real function. It is related to all the terms of the sum in which $\alpha_j \in (-\infty, \infty)\backslash E$. Moreover, in the sum

$$\frac{\sqrt{T(z)} + \sqrt{T(\alpha_j^+)}}{2(z-\alpha_j)\sqrt{T(z)}} + \frac{\sqrt{T(z)} + \sqrt{T(\overline{\alpha}_j^+)}}{2(z-\overline{\alpha}_j)\sqrt{T(z)}}$$
$$= \frac{2z - (\alpha_j + \overline{\alpha}_j)}{2(z-\alpha_j)(z-\overline{\alpha}_j)} + \left(\frac{\sqrt{T(\alpha_j^+)}}{z-\alpha_j} + \frac{\sqrt{T(\overline{\alpha}_j^+)}}{z-\overline{\alpha}_j}\right)\frac{1}{2\sqrt{T(z)}},$$

the first term on the right is a real number for real z. To investigate the second term in $\mathbb{C}_E^+$, we observe that $\sqrt{T(z)}$ takes only pure imaginary values on F^+ and that by the symmetry principle $\sqrt{T(\overline{\alpha}_j^+)} = -\sqrt{T(\alpha_j^+)}$. Hence it follows that for real z the number in parentheses is pure imaginary. At all points of F^+ the square root $\sqrt{T(z)}$ also is a pure imaginary number, and consequently all the sums on the right-hand side of (5.12) are real.

We now integrate the right-hand side of (5.12) over the simple contour B_k $(k = 1, 2, \ldots, g)$ and equate the real part of the integral to zero. If there are points α_j on (a_k, b_k) (there must be an even number) we can deform the contour, in the usual way, close to those points and take the limit. Denoting by $2\beta_k$ the real part of the integral of the first sum over B_k, we obtain the system of equations

$$\sum_{j=1}^{g} \alpha_{jk} d_{j-1} = \beta_k \qquad (k = 1, 2, \dots, g),$$

which is solvable, since, as we showed, $\det \|\alpha_{jk}\|_{j,k=1}^{g} \neq 0$. Setting

$$\Phi(z) = -i \sum_{j=1}^{N} \frac{\sqrt{T(\alpha_j^+)}}{2(z - \alpha_j)\sqrt{T(z)}} + \frac{d_0 + d_1 z + \cdots + d_{g-1} z^{g-1}}{\sqrt{T(z)}} \tag{5.13}$$

and integrating (5.12) from 0 to z on an arbitrary path, we obtain (remembering that $P(0) = 1$)

$$f(z) = \sqrt{P(z)} \exp\left(i \int_0^z \Phi(\tau)\, d\tau\right) \tag{5.14}$$

and correspondingly

$$f_*(z) = \sqrt{P(z)} \exp\left(-i \int_0^z \Phi(\tau) d\tau\right).$$

By the choice of the d_ν ($\nu = 0, 1, \dots, g-1$), we easily obtain

$$\exp\left(i \int_{a_k}^{b_k} \Phi(\tau)\, d\tau\right) = 1, \qquad k = 1, 2, \dots, g.$$

If there are singular points α_j on the interval of integration we avoid them by small semicircles and take limits.

We also show that the real function $\Phi(z)$ is positive on E_u^+. For this purpose we notice that, using (5.12) and (5.12′), we have

$$\frac{f'(z)}{f(z)} - \frac{f_*'(z)}{f_*(z)} = 2i\Phi(z). \tag{5.15}$$

Integrating both sides of (5.15) from 0 to z, we obtain

$$\int_0^z \Phi(\tau)\, d\tau = \frac{1}{2i} \ln \frac{f(z)}{f_*(z)}.$$

Here we suppose that $A(0) = 1$ and, consequently, $f(0) - f_*(0) = 1$. Taking a point $z = x \in E_u^+$ and integrating along the real axis, we obtain

$$\int_0^{x_b} \Phi(\tau)\, d\tau = \frac{1}{2i} \arg \frac{f(x_b)}{f_*(x_b)}. \tag{5.16}$$

Since the modulus $|f(z)/f_*(z)| < 1$ in $\mathbb{C}_E^+$ and equals 1 on E, we have, everywhere on E_u, the inequality

$$\frac{\partial}{\partial y} \ln \left| \frac{f(z)}{f_*(z)} \right| < 0$$

(equality is excluded by Julia's theorem). Hence it follows by the Cauchy–Riemann equations that $\Phi(x) > 0$ throughout E_u^+.

For $A(x)$ and $B(x)$ we obtain the following representation on E:

$$A(x) = \sqrt{P(x)} \cos\left(\int_0^x \Phi(\tau)\, d\tau\right),$$

$$B(x) = \frac{\sqrt{P(x)}}{\sqrt{T(x)}} \sin\left(\int_0^x \Phi(\tau)\, d\tau\right).$$

Therefore, we have solved the problem of constructing $f(z)$ and $f_*(z)$, and hence $A(z)$ and $B(z)$. These formulas can be regarded as generalizations of Markov's formulas for a polynomial positive on $[a, b]$. They give a picture of the oscillatory properties of the pair $A(z)$, $B(z)$. It follows from them that $A(z)$ and $B(z)$ yield a solution of the problem of the polynomial deviating least from zero with a certain weight. These problems will be discussed in more detail in §7.

§6. Application to a problem from the theory of the spectral functions of a string

1. We present some material from the theory of the spectral functions of an inhomogeneous string [5]. A more detailed treatment is given in [6].

On $[0, \infty)$ let there be given an arbitrary nondecreasing right-continuous function $M(x)$ ($M(0) = 0$, $M(\infty) = \infty$). Let $l = l_\infty = \sup\{x \colon M(x) < \infty\}$. We treat $M(x)$ as the mass-distribution function of an inhomogeneous string which is stretched between $x = 0$ and $x = l$ by a single force.

We consider the integral equations

$$\varphi(x; \lambda) = 1 + \lambda \int_{-0}^{x+0} (x - s)\varphi(s; \lambda)\, dM(s),$$

$$\psi(x; \lambda) = x + \lambda \int_{-0}^{x+0} (x - s)\psi(s; \lambda)\, dM(s).$$

The solutions $\varphi(x; \lambda)$ and $\psi(x; \lambda)$ of these equations are fundamental solutions of the homogeneous equation

$$\frac{dy'}{dM(x)} + \lambda y = 0,$$

subject to the boundary conditions

$$\varphi(0; \lambda) = 1, \qquad \varphi'(-0; \lambda) = 0;$$
$$\psi(0; \lambda) = 0, \qquad \psi'(-0; \lambda) = 1.$$

A function $\tilde{\tau}(\lambda)$, nondecreasing on $[0, \infty)$, is called a *positive φ-spectral function of the string* if the mapping $U \colon f \to F$, where $f \in L^2(0, l; dM)$ and

$$F(\lambda) = \int_0^l f(x)\varphi(x; \lambda)\, dM(x),$$

carries $L^2(0, l; dM)$ isometrically to $L^2(0, \infty; d\tilde{\tau})$.

The function

$$\Gamma(\lambda) = \lim_{x\to l} \psi(x;\lambda)/\varphi(x;\lambda)$$

is called the *coefficient of dynamic compliance* of the string; it belongs to the class S, i.e., admits the representation

$$\Gamma(\lambda) = \gamma + \int_0^\infty \frac{d\tau(\lambda)}{t-\lambda} \qquad (\gamma \geq 0,\ d\tau \geq 0).$$

The function $\tau(t)$ determined by this representation is the spectral function of the string and is said to be its *principal spectral function.* The string is uniquely determined by its coefficient of dynamic compliance.

We denote the string with principal spectral function $\tau(\lambda)$ by S_τ.

Let $0 < l_1 < l$ and let $\Gamma_1(\lambda)$ be the coefficient of dynamic compliance of the part of the string corresponding to the interval $[l_1, l]$. Then

$$\Gamma(\lambda) = \frac{\psi'(l_1;\lambda)\Gamma_1(\lambda) + \psi(l_1;\lambda)}{\varphi'(l_1;\lambda)\Gamma_1(\lambda) + \varphi(l_1;\lambda)}. \tag{6.1}$$

If the moments

$$s_k = \int_0^\infty t^k\, d\tau(t), \qquad k = 0, 1, \dots, N-1,$$

exist for the principal spectral function $\tau(t)$ of the string S_τ, an initial section $[0, l_1]$ of S_τ is effectively a weightless filament with $[N/2]$ concentrated masses ("beads"); in this case formula (6.1) becomes formula (3.5):

$$\psi'(l_1;\lambda) = p_0(\lambda), \psi(l_1;\lambda) = p_1(\lambda), \quad \varphi'(l_1;\lambda) = q_0(\lambda), \varphi(l_1;\lambda) = q_1(\lambda),$$

where $q_0(\lambda)$ and $q_1(\lambda)$ are defined by (3.2) and (3.3), and $p_0(\lambda)$ and $p_1(\lambda)$ are the associated polynomials.

2. Let S_{σ_E} be a string with principal spectral function $\sigma_E(t)$, where $\pi d\sigma_E(t) = \sqrt{a(t)/(tb(t))}\,dt$ for $t \in E$ and $d\sigma_E(t) = 0$ for $t \in [0,\infty)\setminus E$. For a polynomial $P(t)$, positive on E, set $d\tau_P(t) = d\sigma_E(t)/P(t)$. For the measure $d\tau_P(t)$ the moments $\{s_k\}_0^{N-1}$ (4.3) exist (we may calculate them from the zeros of $P(t)$, using the function $v(z) = \sqrt{-a(z)/(zb(z))}$, by the method of §2). Hence the initial segment $[0, l_1]$ of S_{τ_P} is a filament with beads. The part of S_{τ_P} that remains after deleting the filament with beads, denoted by S_σ, is described by $\sigma(t)$, the principal spectral function of the remaining part. The following question arises: under what conditions is the string the same (after a shift of l_1) as the string S_{σ_E}?

The answer is given by the following theorem.

THEOREM 6.1. *A necessary and sufficient condition for S_σ to coincide with S_{σ_E} is that the conditions $\varphi'(l_1, b_j) = 0$, $j = 1, 2, \dots, g$, are satisfied, that is, that the polynomial $P(z)$ has a special representation.*

PROOF. Let $\sigma = \sigma_E$. Then (see (4.1) and (4.2))

$$\Gamma_1(\lambda) = \int_0^\infty \frac{d\sigma_E(t)}{t-\lambda} = \frac{1}{\pi}\int_E \frac{\sqrt{a(t)/(t\beta(t))}}{t-\lambda}\,dt = \sqrt{\frac{a(\lambda)}{\lambda b(\lambda)}} = v(\lambda).$$

Consequently,

$$\frac{p_0(\lambda)v(\lambda)+p_1(\lambda)}{q_0(\lambda)v(\lambda)+q_1(\lambda)}=\frac{1}{\pi}\int_E\frac{\sqrt{a(t)/(tb(t))}}{(t-\lambda)P(t)}\,dt.$$

Using the inversion formula (1.1) and the identity (3.4), we find that when $t\in E$,

$$\frac{\sqrt{a(t)/(tb(t))}}{P(t)}=\frac{\sqrt{a(t)/(tb(t))}}{q_0^2(t)a(t)/(tb(t))+q_1^2(t)},$$

from which it follows that $P(t)=q_0^2(t)a(t)/(tb(t))+q_1^2(t)$. Therefore $q_0(t)$ $(=\varphi'(l_1;t))$ is divisible not only by t but by $b(t)$, that is, $\varphi'(l_1;b_j)=0$, $j=1,2,\dots,g$.

Conversely, let $\varphi'(l_1,b_j)=0$, that is, $q_0(b_j)=0$, $j=1,\dots,g$. Then there is a special representation

$$P(z)=A^2(z)+za(z)b(z)B^2(z),$$
$$A(z)=q_1(z),\qquad zb(z)B(z)=q_0(z).$$

We extend the filament with beads S_{σ_E} and see that the principal spectral function $\tau(t)$ of the resulting composite string S_τ coincides with $\tau_P(t)$. In fact,

$$\int_0^\infty\frac{d\tau(t)}{t-\lambda}=\frac{p_0(\lambda)v(\lambda)+p_1(\lambda)}{q_0(\lambda)v(\lambda)+q_1(\lambda)}.$$

Since $q_1(\lambda)/q_0(\lambda)=A(\lambda)/(\lambda b(\lambda)B(\lambda))\in S_+(E)$, $v(\lambda)\in S_+(E)$, and the zeros of $q_0(\lambda)$ lie on E, the denominator of $q_0(\lambda)(v(\lambda)+q_1(\lambda)/q_0(\lambda))$ is not zero and takes real values on $(-\infty,\infty)\backslash E$; therefore, by the inversion formula, $d\tau(t)=0$ on $(-\infty,\infty)\backslash E$. By the same formula, we have on E

$$\begin{aligned}d\tau(t)&=\frac{1}{\pi}\frac{\sqrt{a(t)/(tb(t))}}{q_0^2(t)a(t)/(tb(t))+q_1^2(t)}\,dt\\&=\frac{1}{\pi}\frac{\sqrt{a(t)/(tb(t))}}{ta(t)b(t)B^2(t)+A^2(t)}\,dt\\&=\frac{1}{\pi}\frac{\sqrt{a(t)/(tb(t))}}{P(t)}\,dt,\end{aligned}$$

that is, $d\tau(t)=d\tau_P(t)$, as required.

3. As we noted in the Introduction, the problem that we have discussed is connected with a result of N. I. Akhiezer. Let us explain this in more detail.

It follows from the general results of Gel'fand and Levitan that the function $\sigma_E(\lambda)$ is the spectral function of the Sturm–Liouville problem

$$y''-q_E(x)y+\lambda y=0, \tag{6.2}$$

$$y(0)=1,\qquad y'(0)=h_E \tag{6.3}$$

(with C^∞ coefficients $q_E(x)$). This means that the solution $\Phi(x,\lambda)$ of the system (6.2)–(6.3) generates a generalized Fourier transform

$$\widehat{f}(\lambda)=\int_0^\infty f(x)\Phi(x,\lambda)\,dx, \tag{6.4}$$

which provides a unitary map of $L_2(0,\infty)$ on $L_2(0,\infty; d\sigma_E)$. Here we have the inversion formula

$$f(x) = \int_0^\infty \widehat{f}(\lambda)\Phi(x,\lambda)\,d\sigma_E(\lambda).$$

The convergence of the integral is understood in the mean-square sense with respect to the corresponding measure.

The first deep results on the analytic behavior of the potential $q_E(x)$ were obtained by Akhiezer [7] (see also [8]). Following these investigations, extensions and developments were obtained in several further papers [9]–[11] (also see the survey [12]).

In the case $g = 1$, with the normalization $a_1 = k^2$, $b_1 = 1$ $(0 < k < 1)$, Akhiezer obtained an explicit expression for $q_E(x)$ and h_E, namely

$$q_E(x) = (1-k^2)\frac{cn^2(x,k') - k^2 sn^2(x,k')}{cn^2(x,k') + k^2 sn^2(x,k')}, \qquad h_E = 0 \quad (k'^2 = 1-k^2). \tag{6.5}$$

Moreover, he discovered the following remarkable fact.

For an *arbitrary* polynomial $P(t)$ that is positive on $E = [0,k^2]\cup[1,\infty)$, the spectral function $\tau_P(t)$ defined by the equation $d\tau_P(t) = d\sigma_E(t)/P(t)$ generates a Sturm–Liouville equation with potential $q_E(x+\delta)$, where

$$\delta_P = \frac{1}{2\pi}\int_{E_u} \frac{\ln P(t)}{\sqrt{t(t-k^2)(t-1)}}\,dt. \tag{6.6}$$

The corresponding initial conditions are different from (6.3) and depend in a specific way on the spectral parameter λ. Here the generalized Fourier transform defines an isometric (but not unitary) operator.

To the spectral functions $\sigma_E(t)$ and $\tau_P(t)$ there evidently correspond the same potential if and only if δ_P is a period of $q_E(x)$, i.e.,

$$\delta_P = 2nK' \tag{6.7}$$

with some integer n, where

$$K' = \int_0^1 dt\Big/\sqrt{(1-t^2)(1-k^2t^2)}.$$

The preceding relation agrees with (5.1) in the case $g = 1$, $a_1 = k^2$, $b_1 = 1$, since in this case

$$\int_{a_1}^{b_1} \frac{dx}{\sqrt{|T(x)|}} = \int_{k^2}^{1} \frac{dx}{\sqrt{x(x-k^2)(1-x)}} = 2K'.$$

Consequently, in agreement with Theorem 5.1, a necessary and sufficient condition for the potentials to coincide is that $P(t)$ admits a special representation. When this requirement is satisfied, the boundary conditions for $\Phi_P(x;A)$, defined by the Fourier transform, have the form

$$\Phi_p(0;\lambda) = A(\lambda), \qquad \Phi'_p(0;\lambda) = \lambda b(\lambda)B(\lambda). \tag{6.8}$$

The image under the Fourier transform of the space $L_2(0,\infty)$ is the orthogonal complement in $L_2(0,\infty, d\tau_P)$ of the subspace of polynomials of degree $\leq [N/2]$.

This holds for $d\tau_P(t) = d\sigma_E(t)/P(t)$ and in the case $g > 1$. It provides a simple explanation, and suggests the possibility of extending the isometric operator to a natural unitary operator if we pass from the Sturm–Liouville equation to the string equation.

Notice that the positivity of the spectrum of problem (6.2), (6.3) implies the positivity of the function $\Phi(x;0)$ for $x > 0$. If we set

$$\xi = \int_0^x \frac{dx}{\Phi^2(x;0)}, \qquad \varphi(\xi;\lambda) = \frac{\Phi(x;\lambda)}{\Phi(x;0)}, \qquad \rho(\xi) = \Phi^4(x;0),$$

we can verify by simple transformations that

$$\frac{d^2\varphi}{d\xi^2} + \lambda\rho(\xi)\varphi = 0, \qquad \varphi(0;\lambda) = 1, \quad \varphi'(0;\lambda) = 0.$$

Hence equation (6.2) can be transformed in the usual way into the string equation with mass distribution $M(\xi) = \int_0^\xi \rho(\xi)\,d\xi$. Here it turns out that the φ-spectral function of the string coincides with the Φ-spectral function σ_E of the Sturm–Liouville equation.

If the polynomial $P(t)$ admits a special representation, then the spectral measure $d\tau_P(t) = d\sigma_E(t)/P(t)$ corresponds to a string S_{τ_P} consisting of a filament with beads $\{m_k, l_k\}_1^{[N/2]}$ and a continuous portion S_{τ_P}. Here m_k is the mass of a bead and l_k is the distance between beads m_k and m_{k+1}.

The space of functions with a measure generated by the mass distribution of the string S_{σ_E} is the orthogonal sum of two spaces. One of these coincides with the space $L_2(0, l_\infty; dM)$ generated by the mass distribution on S_{σ_E}; the second is the $[N/2]$-dimensional Hilbert space of complex vectors $\vec{g} = (g_1, g_2, \dots, g_{[N/2]})$ with norm

$$\|\vec{g}\|^2 = \sum_{k=1}^{[N/2]} m_k |g_k|^2.$$

The Fourier transform is a unitary mapping of $L = G \oplus L_2(0, l_\infty; dM)$ on $L_2(0,\infty; d\tau_P)$ $(dM(\xi) = \rho(\xi)\,d\xi)$; if $\widetilde{f} = \vec{g} \oplus f \in \widetilde{L}$ then its Fourier image has the form

$$\widehat{F}(\lambda) = \sum_{k=1}^{[N/2]} g_k m_k \varphi_P(x_k;\lambda) + \int_0^\infty f(\xi)\varphi_P(\xi;\lambda)\,dM(\xi) \tag{6.9}$$

where x_k are the abscissas of the beads, and m_k their masses. Here $\varphi_P(x_k;\lambda)$ are polynomials of degree $k-1$ in λ, orthogonal with weight $\tau_P(\lambda)$ (the numbers l_k, m_k, x_k, and the polynomials $\varphi_P(x_k;\lambda)$ are expressible algebraically in terms of the moments $\{s_k\}_0^{N-1}$ [5]). We have the Parseval equation

$$\int_0^\infty |\widehat{F}(\lambda)|^2\,d\tau_P(\lambda) = \|\vec{g}\|^2 + \int_0^\infty |f(x)|^2\,dM(x)$$

and the inversion formula

$$g_k = \int_0^\infty \varphi_P(x_k;\lambda)\widehat{F}(\lambda)\,d\tau_P(\lambda); \qquad f(\xi) = \int_0^\infty \widehat{F}(\lambda)\varphi_P(\xi,\lambda)\,d\tau_P(\lambda).$$

We observe that now $\varphi_P(\xi;\lambda)$ is a solution of the string equation with initial conditions

$$\varphi(0;\lambda) = A(\lambda), \qquad \varphi'(0;\lambda) = \lambda b(\lambda)B(\lambda),$$

which express the joining conditions between the filament with beads and the string. We can make the inverse transition from the equation of the string (without the filament with beads) to the Sturm–Liouville equation, and correspondingly from $\varphi_P(\xi,\lambda)$ to the solution $\Phi_P(x;\lambda)$ $(= \varphi_P(\xi;\lambda)\Phi(x;0))$ which satisfies (6.8). It is easily seen that the transformation (6.9) becomes the transformation

$$\widehat{F}(\lambda) = \sum_{k=1}^{[N/2]} g_k m_k \varphi_P(x_k;\lambda) + \int_0^\infty F(x)\Phi_P(x;\lambda)\,dx.$$

It is a unitary mapping of $G \oplus L^2(0,\infty)$ on $L^2(0,\infty;d\tau_P)$, so that

$$\int_0^\infty |\widehat{F}(\lambda)|^2\,d\tau_P(\lambda) = \|\vec{g}\|^2 + \int_0^\infty |F(x)|^2\,dx \qquad (g_F = g_f).$$

Therefore the mapping $F \to \int_0^\infty F(x)\Phi_P(x;\lambda)\,dx$, obtained when $\vec{g} = 0$, is an isometric mapping of $L^2(0,\infty)$ to $L^2(0,\infty;d\tau_P)$. More precisely, it is a unitary mapping of $L^2(0,\infty)$ in the orthogonal complement in $L^2(0,\infty;d\tau_P)$ to the linear manifold of polynomials of degree no greater than $N/2$. This was discovered by Akhiezer for the case $g = 1$. However, in his paper he did not raise the question of a complete characterization of the polynomials $A(\lambda)$ and $B(\lambda)$, as was done in this paper for arbitrary g.

On the other hand, we note that when $g = 1$ Akhiezer found the general representation for all positive polynomials on E, and in this way obtained the remarkable result which we discussed in §6.

7. Extremal properties of special pairs of polynomials for a bounded set E

Let us consider the following problem. Let

$$E = [a,b]\setminus \bigcup_{j=1}^{g}(\alpha_j,\beta_j),$$

where $-\infty < a < b < \infty$, and let $P(x)$ be a polynomial of degree $2n$ which is positive on E. To find, among the functions of the form $\psi(x) = X(x)/\sqrt{P(x)}$, where $X(x)$ is a polynomial of degree n with leading coefficient 1, the one deviating least from zero on E. This is, as we said, a generalization of Markov's problem. It is rather easily solved if we suppose that $P(t)$ has the special representation

$$P(x) = A^2(x) + T_E(x)B^2(x). \tag{7.1}$$

We specify more precisely that here

$$T_E(x) = (x-a)(b-x)a(x)b(x),$$

where

$$a(x) = \prod_{j=1}^{g}(x-\alpha_j), \qquad b(x) = \prod_{j=1}^{g}(x-\beta_j),$$

$A(x)$ is a polynomial of degree n, $A(a) > 0$, and $B(x)$ is a polynomial of degree $n-g-1$.

The zeros of $A(x)$ and $(x-a)b(x)B(x)$ are simple, lie on E, and are interlaced. The leading coefficients of $A(x)$ and $B(x)$ are denoted by A_0 and B_0.

THEOREM 7.1. *If the polynomial $P(x)$ of degree $2n$ admits the special representation* (7.1), *then the function $\psi_0(x) = A(x)/(A_0\sqrt{P(x)})$ deviates least from zero on E among all functions $\psi(x) = X(x)/\sqrt{P(x)}$, where $X(x)$ is a polynomial of degree n with leading coefficient* 1. *The minimum deviation is* $1/\sqrt{|A_0|}$.

PROOF. Since

$$\psi_0(x) = \frac{A(x)}{A_0\sqrt{A^2(x) + T_E(x)B^2(x)}},$$

it is evident that the inequality $|\psi_0(x)| \le 1/|A_0|$ holds on E. We denote by ξ_j $(j = 1, 2, \dots, n)$ the zeros of $(x-a)b(x)B(x)$, enumerated in increasing order. Since the zeros of $A(x)$ interlace with ξ_j and $A(a) > 0$, we have $(-1)^{j-1}A(\xi_j) > 0$, so that $\psi_0(\xi_j) = (-1)^{j-1}/A_0$. The required proposition is easily obtained from this. In the first place, it has been established that the deviation of $\psi_0(x)$ from zero is $1/|A_0|$. In the second place, this is the minimum: if for some polynomial $X(x)$ of degree n with leading coefficient 1 the function $\psi(x) = X(x)/\sqrt{P(x)}$ had a smaller deviation, the difference $\psi(x) - \psi_0(x)$, and hence the difference $X(x) - A(x)/A_0$, would change sign at least n times, which is impossible for a polynomial of degree $< n$. In a similar way we can prove the following theorem that describes the extremal property of $B(x)$.

THEOREM 7.2. *If the polynomial $P(x)$ of degree $2n$ admits the special representation* (7.1), *then the function*

$$\chi_0(x) = B(x)\sqrt{T_E(x)/P(x)}/B_0$$

deviates least from zero on E among all functions of the form $\chi(x) = Y(x)\sqrt{T_E(x)/P(x)}$, where $Y(x)$ is a polynomial of degree $n-g-1$ with leading coefficient 1. *The least deviaton is equal to* $1/|B_0|$.

REFERENCES

1. N. I. Akhiezer, *On the Lamé equation*, Khar′kov. Gos. Univ. Uchen. Zap. **138**=Zap. Mekh-Mat. Fak. i Khar′kov. Mat. Obshch. (4) **30** (1964), 5–17. (Russian)

2. A. A. Markov, *Lectures on functions deviating least from zero*, Mimeographed notes, St. Petersburg, 1906; reprinted in his *Selected papers on continued fractions and the theory of functions deviating least from zero*, OGIZ, Moscow, 1948, pp. 244–291. (Russian)

3. M. G. Kreĭn and A. A. Nudel'man, *The Markov moment problem and extremal problems*, "Nauka", Moscow, 1973; English transl., Amer. Math. Soc., Providence, R.I., 1977.

4. N. G. Chebotarev, *Theory of algebraic functions*, OGIZ, Moscow, 1948. (Russian)

5. I. S. Kats and M. G. Kreĭn, *Spectral functions of the string*. Appendix II to the Russian transl. of F. V. Atkinson, *Discrete and continuous boundary problems*, "Mir", Moscow, 1968, pp. 648–737; English transl. in Amer. Math. Soc. Transl. (2) **103** (1973).

6. H. Dym and H. P. McKean, *Gaussian processes, function theory, and the inverse spectral problem*, Academic Press, 1976.

7. N. I. Akhiezer, *Continuous analogs of orthogonal polynomials on a system of intervals*, Dokl. Akad. Nauk SSSR **141** (1961), 263–266; English transl. in Soviet Math. Dokl. 2 (1961).

8. N. I. Akhiezer and I. M. Glazman, *Theory of linear operators in Hilbert space*, 3rd ed., Vol. II, "Vishcha Shkola" (Izdat. Khar'kov. Univ., Kharkov, 1978; English transl. of 1st ed., Ungar, New York, 1961.

9. B. A. Dubrovin, *The inverse problem of scattering theory for periodic finite-zone potentials*, Funktsional. Anal. i Prilozhen. **9** (1975), no. 1, 65–66; English transl. in Functional Anal. Appl. **9** (1975).

10. ——, *A periodic problem for the Korteweg-de Vries equation in the class of finite-zone potentials*, Funktsional. Anal. i Prilozhen **9** (1975), no. 3, 41–51; English transl. in Functional Anal. Appl. **9** (1975).

11. A. R. It·s and V. B. Matveev, *Hill operators with a finite number of gaps*, Funktsional Anal. i Prilozhen. **9** (1975), no. 1, 69–70; English transl. in Functional Anal. Appl. **9** (1975).

12. B. A. Dubrovin, V. B. Matveev, and S. P. Novikov, *Nonlinear equations of Korteweg-de Vries type, finite-zone linear operators, and Abelian varieties*, Uspekhi Mat. Nauk **31** (1976), no. 1 (187), 53–136; English transl. in Russian Math. Surveys **31** (1976).

Translated by RALPH P. BOAS

Amer. Math. Soc. Transl.
(2) Vol. **142**, 1989

Decomposition of Entire Functions of Finite Order into Equivalent Factors(1)

B. N. KHABIBULLIN

The question of decomposing an entire function into factors resembling each other as closely as possible in behavior arose in Krasichkov-Ternovskiĭ's paper [1] in connection with the problem of the local description of closed submodules in spaces of entire functions of exponential type. The result of [1] in this direction was as follows: every entire function $f(z)$ of order 1 whose sequence of zeros has an upper density can be decomposed into the product of two factors, $f(z) = f_1(z) \cdot f_2(z)$, such that $\ln|f_1(z)| - \ln|f_2(z)| = o(|z|)$ outside a set of disks of zero linear density as $z \to +\infty$. Azarin ([2], [3]) extended this result to entire functions of finite order, i.e., he proved that every entire function $f(z)$ of order $\rho > 0$, finite type, can be decomposed into a product $f(z) = f_1(z) \cdot f_2(z)$ such that $\ln|f_1(z)| - \ln|f_2(z)| = o(|z|^\rho)$ outside a set of disks of zero logarithmic density. The problem of decomposing a function into factors with similar behavior at infinity is closely connected with the problem of comparing entire functions according to the closeness of their zeros. In this paper we establish a comparison theorem for canonical products when their zeros differ little from each other (Theorem A), and, as a corollary, Theorem B, which generalizes results of [1] and [3]. We note that the problem of comparing canonical products according to the closeness of their zeros was considered by Levin ([4], pp. 130 and 143 (English pp. 98, 109)) and Krasichkov-Ternovskiĭ [5], and that our proof of Theorem A is essentially based on the results of [5].

---

(1)The author thanks I. F. Krasichkov-Ternovskiĭ for suggesting this problem and supervising his work.

1980 *Mathematics Subject Classification* (1985 *Revision*). Primary 30D20.

Translation of Problems in the Approximation of Functions of Real and Complex Variables (G. V. Ulina, editor), Otdel. Fiz. Mat. Bashkir. Fil. Akad. Nauk SSSR, Ufa, 1983, pp. 161–181; MR **87a**:30048.

§1. Notation and main results

An *E-set* of disks is a union of countably many disks $O_i = \{z\colon |z-h_i| \le \rho_i\}$ $(i = 1, 2, \dots)$ in the complex plane. We suppose that there are only a finite number of disks in any bounded subset of the complex plane.

The variable linear density of an E-set of disks is the function

$$P_E(r) = \frac{1}{r} \sum_{|h_i| \le r} \rho_i \qquad (r > 0).$$

We denote by $O_\rho(x)$ a function of x which is a product of numbers whose moduli are bounded by a number depending only on ρ.

Let there be given two sequences $\Lambda = \{\lambda_i\}$, $\Gamma = \{\gamma_i\}$ of complex numbers $(\lambda_i, \gamma_i \ne 0)$, tending to infinity. We introduce the following products:

$$G(z, p; \Lambda) = \prod_{|\lambda_i| > 0} G(z/\lambda_i, p),$$

$$G^*(z, p; \Lambda) = \prod_{|\lambda_i| \le |z|} G(z/\lambda_i, p-1) \prod_{|\lambda_i| > |z|} G(z/\lambda_i, p),$$

$$G^*(z, p; \Gamma|\Lambda) = \prod_{|\lambda_i| \le |z|} G(z/\gamma_i, p-1) \prod_{|\lambda_i| > |z|} G(z/\gamma_i, p),$$

where $G(u, p) = (1-u)\exp\{u+\dots+u^p/p\}$ is a Weierstrass canonical product.

We denote by $n(t, \Lambda)$ and $n_f(t)$ the number of points of Λ and the number of zeros of $f(z)$ in a closed disk of radius t. We quote a result from [5] (Theorem 4, p. 227), which we shall need in what follows.

THEOREM. *Let the sequences $\Lambda = \{\lambda_i\}$ and $\Gamma = \{\gamma_i\}$ $(\lambda_i, \gamma_i \ne 0,\ i = 1, 2, \dots)$ satisfy the following conditions:*

1) $\sup_{t>0} \frac{n(t, \Lambda\cup\Gamma)}{t^\rho} \le D$ *for some $\rho > 0$,*

2) *there is a number $d \in (0, 1/2)$ such that $|\lambda_i - \gamma_i| \le d|\lambda_i|$ for $i = 1, 2, \dots$.*

Then to each number $\alpha \in (0, 1)$ and $\beta \in (0, 1]$ there corresponds a set of disks E_α with the following properties:

1) *$\Gamma \cup \Lambda \subset E_\alpha$ and each E_α contains at least one point of $\Gamma \cup \Lambda$,*

2) $\overline{\lim}_{r\to\infty} P_{E_\alpha}(r) \le \beta d^{\alpha^2}$,

3) *for $z \notin E_\alpha$ and $p = [\rho]$,*

$$|\ln|G(z, p; \Lambda)| - \ln|G(z, p; \Gamma)|| \le O_\rho(D)\frac{d^{1-\alpha}}{\alpha\beta\sin\pi\alpha}|z|^\rho$$

if ρ is not an integer; and if ρ is an integer

$$|\ln|G^*(z, p; \Lambda)| - \ln|G^*(z, p; \Gamma|\Lambda)|| \le O_\rho(D)\frac{d^{1-\alpha}}{\alpha\beta\sin\pi\alpha}|z|^\rho.$$

However, we can easily see from the method of proof of this theorem ([5], Lemma 4, p. 216, and Lemma 5) that for an appropriate choice of β we may state that

2′) $P_{E_\alpha}(r) = d^{\alpha^2}$ for all $r > 0$,

3′) for $z \notin E_\alpha$ and $p = [\rho]$,

$$\Big|\ln|G(z,p;\Lambda)| - \ln|G(z,p;\Gamma)|\Big| \le O_p(D^2)\frac{d^{1-\alpha}}{\alpha^2 \sin\pi\alpha}|z|^\rho$$

if ρ is not an integer; and if ρ is an integer,

$$\Big|\ln|G^*(z,p;\Lambda)| - \ln|G^*(z,p;\Gamma|\Lambda)|\Big| \le O_p(D^2)\frac{d^{1-\alpha}}{\alpha^2 \sin\pi\alpha}|z|^\rho.$$

We shall prove the following theorems.

THEOREM A. *Let two sequences $\Lambda = \{\lambda_i\}$ and $\Gamma = \{\gamma_i\}$ $(\lambda_j, \gamma_i \neq 0)$ be given, and let the following conditions be satisfied:*

1) $\sup_{t>0} n(t, \Delta\cup\Gamma)/t^\rho < +\infty$ *for some* $\rho > 0$,

2) *for some $\gamma \in (0,1)$ we have $|\lambda_i - \gamma_i| = O(|\lambda_i|/|\lambda_i|^{\rho\gamma})$ as $i \to \infty$.*

Then with every number $\alpha \in (0,1)$ we can associate a set E_α of disks with the following properties:

1) $P_{E_\alpha}(r) = O(1/r^\sigma)$, *where* $\sigma = \alpha^2\rho\gamma$;

2) *for* $z \notin E_\alpha$ *and* $p = [\rho]$

$$\left|\ln|G(z,p;\Lambda)| - \ln|G(z,p;\Gamma)| - \sum_{\rho(1-\gamma)\le k\le\rho} \operatorname{Re}\frac{b_k}{k}z^k\right| \le \frac{\text{const}}{\alpha^2\sin\pi\alpha}|z|^{\rho(1-\gamma+\alpha)},$$

where the constant is independent of α and z, and

$$b_k = \sum_{i=1}^\infty\left(\frac{1}{\lambda_i^k} - \frac{1}{\gamma_i^k}\right).$$

THEOREM B. *Let $f(z)$ be an entire function of order at most ρ, $\rho \in (0,+\infty)$. Then for each $\varepsilon \in (0,1/2)$ the function $f(z)$ can be decomposed into a product of two entire functions, $f(z) = f_1(z)\cdot f_2(z)$, so that the orders of f_1 and f_2 do not exceed ρ and*

$$\big|\ln|f_1(z)| - \ln|f_2(z)|\big| = O(|z|^{\rho(1/2+\varepsilon)})$$

as $z \to \infty$, $z \notin E_\varepsilon$, where E_ε is a set of disks such that $P_{E_\varepsilon}(r) = O(1/r^\sigma)$ (σ is a positive number depending on ρ and ε).

§2. Comparison of canonical products

First we establish two lemmas.

LEMMA 1. *Let there be given two sequences $\Lambda = \{\lambda_i\}$ and $\Gamma = \{\gamma_i\}$ of complex numbers, which for some $\rho > 0$ and $\gamma \in (0,1)$ satisfy the following conditions:*

1) $\lambda_i, \gamma_i \neq 0$, $i = 1, 2, \dots$, $\sup_{t>0}\frac{n(t,\Lambda\cup\Gamma)}{t^\rho} \le D$,

2) $|\lambda_i - \gamma_i| \le c|\lambda_i|/|\lambda_i|^{\rho\gamma}$ *for every* i $(c > 1)$.

Then we have, for $t > (2c)^{1/(\rho\gamma)}$,

1) *for* $\rho(1-\gamma) < k \le \rho$ (*k an integer*) *the inequality*

$$\sum_{|\lambda_i|>t} \left| \frac{1}{\lambda_i^k} - \frac{1}{\gamma_i^k} \right| \le O_\rho(cD) t^{\rho(1-\gamma)-k} \tag{1}$$

is satisfied,

2) *for* $k = \rho(1-\gamma)$ (*that is, if* $\rho(1-\gamma)$ *is an integer*)

$$\sum_{(2c)^{1/(\rho\gamma)} < |\lambda_i| \le t} \left| \frac{1}{\lambda_i^k} - \frac{1}{\gamma_i^k} \right| \le O_\rho(cD)(1+\ln t), \tag{2}$$

3) *for* $k < \rho(1-\gamma)$

$$\sum_{(2c)^{1/(\rho\gamma)} < |\lambda_i| \le t} \left| \frac{1}{\lambda_i^k} - \frac{1}{\gamma_i^k} \right| \le O_\rho(cD) t^{\rho(1-\gamma)-k}. \tag{3}$$

PROOF. It follows from condition 2) that

$$\left(1 - \frac{c}{|\lambda_i|^{\rho\gamma}}\right) |\lambda_i| \le |\gamma_i| \le \left(1 + \frac{c}{|\lambda_i|^{\rho\gamma}}\right) |\lambda_i|,$$

that is, for $|\lambda_i| > (2c)^{1/(\rho\gamma)}$ we obtain

$$\frac{1}{2}|\lambda_i| < |\gamma_i| < \frac{3}{2}|\lambda_i|. \tag{4}$$

Hence, for $|\lambda_i| > (2c)^{1/(\rho\gamma)}$ we obtain

$$\begin{aligned} \left| \frac{1}{\lambda_i^k} - \frac{1}{\gamma_i^k} \right| &= \frac{|\lambda_i - \gamma_i| |\lambda_i^{k-1} + \cdots + \gamma_i^{k-1}|}{|\lambda_i|^k |\gamma_i^k|} \\ &\le O_\rho(c) \frac{|\lambda_i|^{k-1} |\lambda_i|}{|\lambda_i^{\rho\gamma}| |\lambda_i|^{2k}} \\ &\le O_\rho(c) \frac{1}{|\lambda_i|^{k+\rho\gamma}}. \end{aligned}$$

Hence, for $t > (2c)^{1/(\rho\gamma)}$ and $\rho(1-\gamma) < k \le \rho$,

$$\sum_{|\lambda_i|>t} \left| \frac{1}{\lambda_i^k} - \frac{1}{\gamma_i^k} \right| \le O_\rho(c) \int_t^\infty \frac{1}{t^{k+\rho\gamma}} \, dn(t, \Lambda).$$

Integrating by parts, we obtain (1).

For $k = \rho(1-\gamma)$ we have, by (4),

$$\begin{aligned} \sum_{(2c)^{1/(\rho\gamma)} \le |\lambda_i| \le t} \left| \frac{1}{\lambda_i^k} - \frac{1}{\gamma_i^k} \right| &\le O_\rho(c) \sum_{1 < |\lambda_i| \le t} \frac{1}{|\lambda_i|^{k+\rho\gamma}} \\ &\le O_\rho(cD)(1+\ln t). \end{aligned}$$

Inequality (3) is obtained similarly.

LEMMA 2. *Let there be given two sequences satisfying the hypotheses of Lemma* 1.

Then to every $\alpha \in (0, 1)$ *and every* $t > 2^{1/\gamma}(32c)^{1/(\rho\gamma)}$ *there corresponds a set of disks* $E_{t,\alpha}$ *with the properties*:

1) $P_{E_{t,\alpha}}(r) \le 2^{\rho}c/t^{\sigma}$ *for arbitrary* $r > 0$, *where* $\sigma = \alpha^2\rho\gamma$;

2) *for* $p = [\rho]$, $t < |z| < 4t$, *and* $z \notin E_{t,\alpha}$,

$$\left|\ln|G(z, p; \Lambda)| - \ln|G(z, p; \Gamma)| - \operatorname{Re}\sum_{\rho(1-\gamma)<k\le\rho} \frac{b_k}{k} z^k\right| \le \frac{\text{const}}{\alpha^2 \sin\pi\alpha}|z|^{\rho(1-\gamma+\alpha)}, \tag{5}$$

where

$$b_k = \sum_{i=1}^{\infty}\left(\frac{1}{\lambda_i^k} - \frac{1}{\gamma_i^k}\right)$$

and the constant depends only on ρ, c, D, *and* γ.

PROOF. We give the proof under the assumption that ρ is an integer. For a nonintegral ρ the proof is considerably simpler.

Notice that by Lemma 1 the series

$$\sum_{i=1}^{\infty}\left(\frac{1}{\lambda_i^k} - \frac{1}{\gamma_i^k}\right)$$

converges for $\rho|1-\gamma| < k \le \rho$.

We denote by Λ_t the set $\{\lambda_i\colon |\lambda_i| > \frac{t}{2}\}$, and by Γ_t the set $\{\gamma_i\colon |\gamma_i| > \frac{t}{2}\}$.

Supposing that ρ is integral, we represent the left-hand side of (5) in the following form:

$$\begin{aligned}
&\ln|G(z, p; \Lambda)| - \ln|G(z, p; \Gamma)| - \operatorname{Re}\sum_{\rho(1-\gamma)<k\le\rho}\frac{b_k}{k}z^k \\
&= \ln|G(z, [\rho(1-\gamma)]; \Lambda - \Lambda_t)| - \ln|G(z, [\rho(1-\gamma)]; \Gamma \setminus \Gamma_t)| \\
&\quad + \operatorname{Re}\sum_{\rho(1-\gamma)<k\le\rho}\ \sum_{|\lambda_i|\le\frac{1}{2}t}\frac{z^k}{k}\left(\frac{1}{\lambda_i^k} - \frac{1}{\gamma_i^k}\right) + \ln|G^*(z, p; \Lambda_t)| \\
&\quad - \ln|G^*(z, p; \Gamma_t|\Lambda_t)| + \operatorname{Re}\sum_{\frac{1}{2}t<|\lambda_i|\le|z|}\frac{z^{\rho}}{\rho}\left(\frac{1}{\lambda_i^{\rho}} - \frac{1}{\gamma_i^{\rho}}\right) - \operatorname{Re}\sum_{\rho(1-\gamma)<k\le\rho}\frac{b_k}{k}z^k \\
&= (\ln|G(z[\rho(1-\gamma)]; \Lambda\setminus\Lambda_t)| - \ln|G(z, [\rho(1-\gamma)]; \Gamma\setminus\Gamma_t)|) \\
&\quad + \ln|G^*(z, p; \Lambda_t)| - \ln|G^*(z, p; \Gamma_t|\Lambda_t)| \\
&\quad + \operatorname{Re}\sum_{\rho(1-\gamma)<k\le\rho}\frac{z^k}{k}\sum_{|\lambda_i|>\frac{1}{2}t}\left(\frac{1}{\lambda_i^k} - \frac{1}{\gamma_i^k}\right) + \operatorname{Re}\frac{z^{\rho}}{\rho}\sum_{|\lambda_i|>|z|}\left(\frac{1}{\lambda_i^{\rho}} - \frac{1}{\gamma_i^{\rho}}\right).
\end{aligned}$$

By Lemma 1, the last two sums can be estimated, for $t > (2c)^{1/(\rho\gamma)}$ and $|z| \le 4t$, by $O_\rho(cD)|z|^{\rho(1-\gamma)}$.

$$
\begin{aligned}
&\ln|G(z,[\rho(1-\gamma)];\Lambda\setminus\Lambda_t)| - \ln|G(z,[\rho(1-\gamma)];\Gamma\setminus\Gamma_t)| \\
&\quad = \sum_{|\lambda_i|<\frac{1}{2}t}\left(\ln\left|1-\frac{z}{\lambda_i}\right| - \ln\left|1-\frac{z}{\gamma_i}\right|\right) + \sum_{k\le\rho(1-\gamma)}\sum_{|\lambda_i|<\frac{1}{2}t}\operatorname{Re}\frac{z^k}{k}\left(\frac{1}{\lambda_i^k}-\frac{1}{\gamma_i^k}\right) \\
&\quad = \sum_{(32c)^{1/(\rho\gamma)}<|\lambda_i|\le\frac{1}{2}t}\ln\left|1+z\frac{\lambda_i-\gamma_i}{\lambda_i(\gamma_i-z)}\right| + \sum_{|\lambda_i|\le(32c)^{1/(\rho\gamma)}}\left(\ln\left|1-\frac{z}{\lambda_i}\right| - \ln\left|1-\frac{z}{\gamma_i}\right|\right) \\
&\qquad + \sum_{k\le\rho(1-\gamma)}\sum_{|\lambda_i|\le(2c)^{1/(\rho\gamma)}}\operatorname{Re}\frac{z^k}{k}\left(\frac{1}{\lambda_i^k}-\frac{1}{\gamma_i^k}\right) \\
&\qquad + \sum_{k\le\rho(1-\gamma)}\sum_{(2c)^{1/(\rho\gamma)}<|\lambda_i|\le\frac{1}{2}t}\operatorname{Re}\frac{z^k}{k}\left(\frac{1}{\lambda_i^k}-\frac{1}{\gamma_i^k}\right).
\end{aligned}
\tag{6}
$$

It is easily seen that, for $t < |z| \le 4t$,

$$
\begin{aligned}
\sum_{(32c)^{1/(\rho\gamma)}<|\lambda_i|\le\frac{1}{2}t}\left|\ln\left|1+z\frac{\lambda_i-\gamma_i}{\lambda_i(\gamma_i-z)}\right|\right| &\le \sum_{(32c)^{1/(\rho\gamma)}<|\lambda_i|\le\frac{1}{2}t}|z|\frac{c}{|\lambda_i|^{\rho\gamma}|\gamma_i-z|} \\
&\le \sum_{(32c)^{1/(\rho\gamma)}<|\lambda_i|\le\frac{1}{2}t}\frac{16c}{|\lambda_i|^{\rho\gamma}} \le O_\rho(cD)|z|^{\rho(1-\gamma)}.
\end{aligned}
$$

The second sum in (6) can be estimated by $\operatorname{const}\ln|z|$, where the constant depends only on ρ, γ, and c. By Lemma 1, the last sum in (6) can be estimated by $O_\rho(cD)|z|^{\rho(1-\gamma)}\ln|z|$, and the penultimate sum by $\operatorname{const}|z|^{\rho(1-\gamma)}$, where the constant depends only on ρ, γ, and c.

To complete the proof of Lemma 2, it remains only to estimate

$$\ln|G^*(z,p;\Lambda_t)| - \ln|G^*(z,p;\Gamma_t|\Lambda_t)|.$$

We apply the theorem stated in §1, making use of some remarks.

For $|\lambda_i| > \frac{t}{2}$, by the hypotheses of the lemma,

$$|\lambda_i-\gamma_i| \le \frac{c}{|\lambda_i|^{\rho\gamma}}|\lambda_i| \le \frac{2^\rho c}{t^{\rho\gamma}}|\lambda_i|,$$

and for $t > 2^{1/\gamma}(32c)^{1/(\rho\gamma)}$ we obtain $2^\rho c/t^{\rho\gamma} \le 1/2$.

Applying the theorem of §1 with $d = 2^\rho c/t^{\rho\gamma}$, we find that with every $\alpha\in(0,1)$ we can associate a set $E_{t,\alpha}$ of disks with the properties:

1) $\Gamma_t\cup\Lambda_t\subset E_{t,\alpha}$ and each disk $E_{t,\alpha}$ contains at least one point in $\Gamma_t\cup\Lambda_t$;

2′) $P_{E_{t,\alpha}}(r) \le (2^\rho c/t^{\rho\gamma})^{d^2} \le 2^\rho c/t^\sigma$;

3′) for $z \notin E_{t,\alpha}$ and $|z| \le 4t$, $p = [\rho]$,

$$\begin{aligned} &|\ln|G^*(z, p; \Lambda_t)| - \ln|G^*(z, p; \Gamma_t|\Lambda_t)|| \\ &\le O_\rho(D^2)\frac{(2^\rho c)^{1-\alpha}|z|^\rho}{t^{\rho(1+\alpha)}\alpha^2 \sin \pi\alpha} \\ &\le O_\rho(cD^2)\frac{|z|^{\rho(1-\gamma+\alpha)}}{\alpha^2 \sin \pi\alpha}. \end{aligned}$$

Hence Lemma 2 is established.

Theorem A. *Let there be given two sequences $\Lambda = \{\lambda_i\}$ and $\Gamma = \{\gamma_i\}$ ($\gamma_i, \lambda_i \ne 0, i = 1, 2, \dots$) of complex numbers, and let the following conditions be satisfied*:

1) $\sup_{t>0} \frac{n(t,\Lambda\cup\Gamma)}{t^\rho} < +\infty$ (*ρ is a given positive number*),

2) $|\lambda_i - \gamma_i| = O(1/|\lambda_i|^{\rho\gamma})|\lambda_i|$ *as $i \to \infty$, for some $\gamma \in (0, 1)$.*

Then with every number $\alpha \in (0, 1)$ we can associate a set of disks E_α with the properties:

1) $P_{E_\alpha} = O(1/r^\sigma)$ *as $r \to \infty$, where $\sigma = \alpha^2\rho\gamma$,*

2) *for $z \notin E_\alpha$ and $p = [\rho]$*

$$\left|\ln|G(z, p; \Lambda)| - \ln|G(z, p; \Gamma)| - \sum_{\rho(1-\gamma)<k\le\rho} \operatorname{Re}\frac{b_k}{k}z^k\right| \le \frac{\text{const}}{\alpha^2 \sin \pi\alpha}|z|^{\rho(1-\gamma+\alpha)},$$

where the constant is independent of α and z, and b_k is given by (5).

Proof. By the hypotheses of the theorem there are numbers $D > 0$ and $c > 0$ such that $n(t, \Delta\cup\Gamma)/t^\rho \le D$ for arbitrary t and $|\lambda_i - \gamma_i| \le c|\lambda_i|/|\lambda_i|^{\rho\gamma}$ for every i. Let $E_\alpha^{(n)}$ be the set of disks $E_{2^n,\alpha}$ introduced in Lemma 2, with $n > \log_2 2^{1/\gamma}(32c)^{1/(\rho\gamma)}$.

By Lemma 2,

$$P_{E_\alpha^{(n)}}(r) \le 2^\rho c/2^{n\sigma} < 1/8$$

for $n \ge n_0 = [\log_2 2^{1/\gamma}(32c)^{1/(\rho\gamma)}] + 1$.

We divide the plane into annuli $u_n = \{z\colon 2^n < |z| \le 2^{n+1}\}$ for $n > n_0$. Since $P_{E_\alpha^{(n)}}(r) < \frac{1}{8}$, in each annulus u_n we can find a circle $|z| = t_n$ such that $E_\alpha^{(n)} \cup E_\alpha^{(n-1)}$ is disjoint from this circle.

By construction, $t_n < t_{n+1} \le 4t_n$. In each annulus $S_n = \{z\colon t_n < |z| \le t_{n+1}\}$ there are only disks from the set $E_\alpha^{(n)}$. We introduce the new set of disks

$$E_\alpha = \{z\colon |z| \le t_{n_0+1}\} \cup \left(\bigcup_{n\ge n_0+1} (E_\alpha^{(n)} \cap S_n)\right).$$

Let us estimate its variable density. For $n > n_0$, by Lemma 2,

$$\sum_{h_i\in E_\alpha^{(n)}\cap S_n} \rho_i \le \frac{2^\rho c}{2^{n\sigma}}t_{n+1} \le c\cdot 2^{\rho+2}\cdot 2^{n(1-\gamma)}. \tag{7}$$

Let $t_n < r \le t_{n+1}$. Then by (7)

$$\begin{aligned} P_{E_\alpha}(r) &= \frac{1}{r}\sum_{|h_i|\le r}\rho_i \le \frac{1}{\rho}\left(\sum_{k=n_0+1}^{n}\ \sum_{h_i\in E_\alpha^{(t)}\cap S_k}\rho_i + t_{n_0+1}\right) \\ &\le \frac{1}{r}\left(\sum_{k=n_0+1}^{n} c2^{\rho+1}2^{k(1-\sigma)} + 2^{n_0+2}\right) \\ &\le \frac{1}{r}\left(2^{\rho+2}c\frac{2^{(n_0+1)(1-\sigma)}(2^{(n-n_0)(1-\sigma)}-1)}{1-2^{1-\sigma}} + 2^{n_0+2}\right) \\ &\le \text{const}\,\frac{1}{r}\cdot\frac{2^n}{2^{(n+2)\sigma}} \le \text{const}\,\frac{1}{r^\sigma}, \end{aligned}$$

where the constant depends only on ρ, c, σ, and n_0. Let $t_n < |z| \le t_{n+1}$ and $z \notin E_\alpha$; then $2^n < |z| < 4\cdot 2^n$ and $z \notin E_\alpha^{(n)}$. Consequently, by Lemma 2, we obtain the inequality

$$\left|\ln|G(z,p;\Lambda)| - \ln|G(z,p;\Gamma)| - \operatorname{Re}\sum_{\rho(1-\gamma)<k\le\rho}\frac{b_k}{k}z^k\right| \le \frac{\text{const}}{\alpha^2\sin\pi\alpha}|z|^{\rho(1-\gamma+\alpha)},$$

where the constant depends only on ρ, c, D, and γ. This completes the proof of the theorem.

§3. Decomposition of an entire function of finite order into equivalent factors

The following lemma is easily obtained from a known theorem on a lower bound for the modulus of a holomorphic function ([4], p. 33 (English p. 21)).

LEMMA 3. *Let $f(z)$ be an entire function of order $\rho > 0$ and normal type. Then there is a set E of disks such that*

1) *$rP_E(r) = O(1)$, i.e., the sum of the radii of the disks in E is bounded,*

2) *for $z \notin E$ and for some $c > 0$ we have the inequality*

$$\ln|f(z)| > -c|z|^\rho \ln|z|.$$

LEMMA 4. *Let $f(z)$ be an entire function of order ρ and $\overline{\lim}_{t\to\infty} n_f(t)/t^\rho < \infty$. If $\gamma \in (0,1)$, then $f(z)$ can be decomposed into the product of three entire functions, $f(z) = f_1(z)f_2(z)f_3(z)$, so that*

1) *$f_3(z)$ has order not exceeding $2\rho\gamma$, and*

$$\overline{\lim_{t\to\infty}}\, n_{f_3}(t)/t^{2\rho\gamma} < \infty,$$

2) *with each $\alpha \in (0,1)$ we can associate a set $E_{\alpha,\gamma}$ of disks such that, for $z \notin E_{\alpha,\gamma}$,*

$$|\ln|f_1(z)| - \ln|f_2(z)|| \le \frac{\text{const}}{\alpha^2\sin\pi\alpha}|z|^{\rho(1-\gamma+\alpha)},$$

where the constant is independent of α, *and* $P_{E_{\alpha,\gamma}}(r) = O(1/r^\sigma)$, *where* $\sigma = \alpha^2\rho\gamma$.

PROOF. We divide the complex plane into annuli $S_n = \{z\colon n^{1/(\rho\gamma)} < |z| \le (n+1)^{1/(\rho\gamma)}\}$ $(n = 0, 1, \dots)$, and we divide each S_n, $n > 0$, by rays $\arg z = 2\pi k/n$ $(k = 0, 1, \dots, n-1)$ into cells

$$\Delta_{n,k} = \{z\colon n^{1/(\rho\gamma)} < |z| \le (n+1)^{1/(\rho\gamma)}, \frac{2\pi k}{n} \le \arg z < \frac{2\pi(k+1)}{n}, k = 0, 1, \dots, n-1\}.$$

The distance between the circular boundaries of S_n is

$$(n+1)^{1/(\rho\gamma)} - n^{1/(\rho\gamma)} = n^{1/(\rho\gamma)}((1+n^{-1})^{1/(\rho\gamma)} - 1) \le \text{const}\cdot n^{1/(\rho\gamma)-1},$$

where the constant is independent of n. If we take $r_n = n^{1/(\rho\gamma)}$, then

$$r_{n+1} - r_n < \text{const}\,\frac{n^{1/(\rho\gamma)}}{(n^{1/(\rho\gamma)})^{\rho\gamma}} = \text{const}\,\frac{r_n}{r_n^{\rho\gamma}}.$$

The large arc of the cell $\Delta_{n,k}$ is equal to

$$\frac{2\pi(n+1)^{1/(\rho\gamma)}}{n} = \frac{2\pi(1+n^{-1})^{1/(\rho\gamma)} n^{1/(\rho\gamma)}}{n} \le \text{const}\,\frac{r_n}{r_n^{\rho\gamma}},$$

where the constant is independent of n.

Hence the diameter of $\Delta_{n,k}$ does not exceed

$$\text{const}\,\frac{r_n}{r_n^{\rho\gamma}} = O\left(\frac{r_n}{r_{n+1}^{\rho\gamma}}\right) \quad \text{as } n \to \infty. \tag{9}$$

Let A be the sequence of zeros of the entire function $f(z)$. We separate A into groups of points $A_{n,k} = A \cap \Delta_{n,k}$, and then divide each $A_{n,k}$ into three parts so that the following conditions are satisfied: if $A_{n,k}$ contains an even number of points then half of them form the group $\Delta_{n,k}$, the other half form the group $\Gamma_{n,k}$, and $K_{n,k} = \{\varnothing\}$; if $A_{n,k}$ contains an odd number $2m+1$ of points, then $\Lambda_{n,k}$ and $\Gamma_{n,k}$ each contain m points, and $K_{n,k}$ contains a single point. Since each cell $\Delta_{n,k}$ contains the same number of points of the sequence

$$\Lambda = \bigcup_{n\ge 0}\left(\bigcup_{k=0}^{n-1} \Lambda_{n,k}\right)$$

and of the sequence

$$\Gamma = \bigcup_{n\ge 0}\left(\bigcup_{k=0}^{n-1} \Gamma_{n,k}\right),$$

these sequences can be indexed so that if $\lambda_i \in \Lambda$ and $\lambda_i \in \Delta_{n,k}$, then $\gamma_i \in \Delta_{n,k}$, where γ_i is a point of Γ with the same index as λ_i. According to (9), for $\lambda_i \in \Delta_{n,k}$ we have the inequality

$$|\lambda_i - \gamma_i| \le \text{const}\, r_n/r_{n+1}^{\rho\gamma} = \text{const}\,|\lambda_i|/|\lambda_i|^{\rho\gamma},$$

where the constants are independent of i. Therefore $\lambda_i, \gamma_i \neq 0$ for all i, $|\lambda_i - \gamma_i| = O(1/|\lambda_i|^{\rho\gamma})|\lambda_i|$ as $i \to \infty$, and $\sup n(t, \Lambda \cup \Gamma)/t^\rho < \infty$.

Then, by Theorem A, with each point $\alpha \in (0, 1)$ we can associate a set of disks $E_{\alpha,\gamma}$ with the properties:

1) $P_{E_{\alpha,\gamma}}(r) = O(1/r^\sigma)$ as $r \to \infty$, where $\sigma = \alpha^2\rho\gamma$,

2) for $z \notin E_{\alpha,\gamma}$ and $p = [\rho]$

$$\left| \ln|G(z, p; \Lambda)| - \ln|G(z, p; \Gamma)| - \sum_{\rho(1-\gamma)<k\leq\rho} \operatorname{Re}\frac{b_k}{k} z^k \right| \leq \frac{\text{const}}{\alpha^2 \sin\pi\alpha} |z|^{\rho(1-\gamma+\alpha)}, \tag{10}$$

where the constant is independent of α and z.

We now estimate the upper density of the set

$$K = \bigcup_{n\geq 0}\left(\bigcup_{k=0}^{n-1} K_{n,k}\right).$$

Let $r_n < t \leq r_{n+1}$. By the construction of K, each annulus S_n contains at most $n + 1$ points of K. Consequently if we denote the points of K by x_i, then for $n \geq 1$

$$h(t, K) = \sum_{|x_i|<t} 1 = \sum_{k=0}^{n} \sum_{x_i\in S_k} 1 \leq \sum_{k=0}^{n} (k+1) < 3n^2 \leq 3r_n^{2\rho\gamma} < 3t^{2\rho\gamma}.$$

Since, by the construction, $K \subset \Lambda \cup \Gamma$, we find that

$$\overline{\lim_{t\to\infty}}\, n(t, K)/t^{\rho_1} < \infty,$$

where $\rho_1 = \min\{2\rho\gamma, p\}$. Hence it follows that the canonical product $G(z, [\rho_1]; K)$ converges and represents an entire function of order at most $\min\{2\rho\gamma, p\}$.

It is clear that

$$\frac{f(z)}{G(z, p; \Lambda)G(z, p; \Gamma)G(z, [\rho_1]; K)} = z^m e^{Q(z)},$$

where $Q(z)$ is a polynomial of degree at most p. If $P(z)$ denotes the polynomial $\sum_{\rho(1-\gamma)<k\leq\rho}(b_k z^k/k)$ in (10), then

$$f(z) = (z^m G(z, p; \Lambda) \exp\{-\tfrac{1}{2}(P(z) - Q(z))\}) \times \theta(z, p; \Gamma) \cdot \exp\{\tfrac{1}{2}(P(z) + Q(z))\} G(z, [\rho_1]; k).$$

Set

$$f_1(z) = z^m G(z, p; \Lambda) \exp\{-\tfrac{1}{2}(P(z) - Q(z))\},$$
$$f_2(z) = G(z, p; \Gamma) \exp\{\tfrac{1}{2}(P(z) + Q(z))\},$$
$$f_3(z) = G(z, [\rho_1]; K).$$

Then, according to (10), the lemma is established.

THEOREM B. *Let $f(z)$ be an entire function of order at most $\rho > 0$. Then for each $\varepsilon \in (0, 1/2)$ we can decompose $f(z)$ into the product of two entire functions $f(z) = f_1(z) \cdot f_2(z)$ such that $\ln|f_1(z)| - \ln|f_2(z)| = O(|z|^{(1/2+\varepsilon)\rho})$ as $z \to \infty$ outside an exceptional set E_t of disks such that $P_{E_t}(r) = O(1/r^\sigma)$ as $z \to \infty$, where $\sigma > 0$ and σ is independent of ε.*

PROOF. First suppose that $\sup_{t>0} n_f(t)/t^\rho < \infty$. Then we can apply Lemma 4 with $\gamma_0 = \frac{(1-\varepsilon)}{2}$ and $\alpha_0 = \varepsilon/2$. According to this lemma $f(z)$ can be represented in the form $f(z) = f_1^{(1)}(z) f_2^{(1)}(z) f_3^{(1)}(z)$, where the order of $f_3^{(1)}(z)$ is at most $2\rho\gamma_0 = (1-\varepsilon)\rho$, and

$$|\ln|f_1^{(1)}(z)| - \ln|f_2^{(1)}(z)|| \le \frac{\text{const}}{\varepsilon^2}|z|^{\rho(\frac{1}{2}+\varepsilon)}$$

outside a set E_1 of disks such that $P_{E_1}(r) = O(1/r^{\sigma_1})$, where $\sigma_1 = \alpha_0^2\rho\gamma_0 = \frac{1}{8}\varepsilon^2(1-\varepsilon)\rho$. Moreover, by Lemma 4,

$$\varlimsup_{t\to\infty} \frac{n_{f_3^{(1)}}(t)}{t^{(1-\varepsilon)\rho}} < \infty.$$

Consequently Lemma 4 is applicable to $f_3^{(1)}(z)$ with $\gamma = \gamma_0 - \varepsilon$ and $\alpha = \alpha_0$, i.e., $f_3^{(1)}(z) = f_1^{(2)}(z) f_2^{(2)}(z) f_3^{(2)}(z)$, where

$$\varlimsup_{t\to\infty} \frac{n_{f_3^{(2)}}(t)}{t^{(1-\varepsilon)^2\rho}} < +\infty \quad \text{and} \quad |\ln|f_1^{(2)}(z)| - \ln|f_2^{(2)}(z)|| \le \frac{\text{const}}{\varepsilon^2}|z|^{\rho(1-\varepsilon)(\frac{1}{2}+\varepsilon)}$$

outside a set E_2 of disks such that $P_{E_2}(r) = O(1/r^{\sigma_2})$, where $\sigma_2 = \frac{1}{8}\varepsilon^2(1-\varepsilon)^2\rho$. Similarly, we apply Lemma 4, with $\gamma = \gamma_0 - 2\varepsilon$ and $\alpha = \alpha_0$, to $f_3^{(2)}(z)$, $f_3^{(3)}(z)$, and so on. At the kth step we will obtain $f_3^{(k-1)}(z) = f_1^{(k)}(z) f_2^{(k)}(z) f_3^{(k)}(z)$, where the entire function is of order at most $(1-\varepsilon)^k\rho$ and

$$\varlimsup_{t\to\infty} \frac{n_{f_3^{(k)}}(t)}{t^{(1-\varepsilon)^k\rho}} < \infty;$$

also

$$|\ln|f_1^{(k)}(z)| - \ln|f_2^{(k)}(z)|| \le \frac{\text{const}}{\varepsilon^2}|z|^{\rho(1-\varepsilon)^{k-1}(\frac{1}{2}+\varepsilon)}$$

outside a set E_k of disks such that

$$P_{E_k}(r) = O(1/r^{\sigma_k}), \quad \text{where } \sigma_k = \tfrac{1}{8}\varepsilon^2(1-\varepsilon)^k.$$

With $k_0 = [2/|\ln(1-\varepsilon)|] + 1$ the order of $f_3^{(k_0)}(z)$ is less than $\rho/2$, and therefore by Lemma 3 we have, outside a set E_{k_0+1} for $f_3^{(k_0)}(z)$, $\ln|f_3^{(k_0)}(z)| = O(|z|^{\rho/2})$ as $z \to \infty$; moreover, $rP_{E_{k_0+1}}(r) = O(1)$ as $r \to \infty$.

Set

$$f_1(z) = \prod_{k=1}^{k_0} f_1^{(k)}(z), \qquad f_2(z) = \left(\prod_{k=1}^{k_0} f_2^{(k)}(z)\right) f_3^{(k_0)}(z).$$

Then, according to the preceding discussion, outside the set of disks $E_\varepsilon = \bigcup_{k=1}^{k_0+1} E_k$ we have

$$\ln|f_1(z)| - \ln|f_2(z)| = O(|z|^{(\frac{1}{2}+\varepsilon)\rho}) \quad \text{as } z \to \infty \ (z \notin E);$$

moreover, $P_{E_\varepsilon}(r) = O(1/r^\sigma)$, where $\sigma = \frac{1}{8}\varepsilon^2(1-\varepsilon)^{k_0}\rho$.

Let $f(z)$ be any entire function of order ρ; then

$$\overline{\lim}\, n_f(t)/t^{\rho(1+\frac{\varepsilon}{2})} < \infty$$

and, according to the preceding discussion, $f(z) = f_1(z) \cdot f_2(z)$, with

$$\ln|f_1(z)| - \ln|f_2(z)| = O(|z|^{(\rho+\frac{1}{2}\varepsilon\rho)(\frac{1}{2}+\frac{1}{2}\varepsilon)}) = O(|z|^{\rho(\frac{1}{2}+\varepsilon)})$$

for $z \notin E_\varepsilon$, $z \to \infty$, where E_ε is a set of disks such that $P_{E_\varepsilon}(r) = O(1/r^\sigma)$, and σ is a positive number depending on ε. This completes the proof of the theorem.

REMARK. By the construction of $f_1(z)$ and $f_2(z)$ it is easy to see (see Lemma 4) that the orders of $f_1(z)$ and $f_2(z)$ do not exceed ρ.

REFERENCES

1. I. F. Krasichkov-Ternovskiĭ, *Invariant subspaces of analytic functions.* I: *Spectral synthesis on convex domains*, Mat. Sb. **87(129)** (1972), 459–489; English transl. in Math. USSR Sb. **16** (1972).

2. V. S. Azarin, *On the rays of completely regular growth of an entire function*, Mat. Sb. **79(121)** (1969), 463–476; English transl. in Math. USSR Sb. **8** (1969).

3. ——, *On the decomposition of an entire function of finite order into components of prescribed growth*, Mat. Sb. **90(132)** (1973), 229–230; English transl. in Math. USSR Sb. **19** (1973).

4. B. Ya. Levin, *Distribution of zeros of entire functions*, GITTL, Moscow, 1956; English transl., Amer. Math. Soc., Providence, R.I., 1964.

5. I. F. Krasichkov-Ternovskiĭ, *Comparison of entire functions of finite order according to the distribution of their zeros*, Mat. Sb. **70(112)** (1966), 198–230. (Russian)

Translated by R. P. BOAS

Amer. Math. Soc. Transl.
(2) Vol. **142**, 1989

On Coincidence of the Ideal of Strictly Singular Operators with the Ideal of Φ-Admissible Perturbations in Lorentz Spaces

G. I. RUSSU

One of the most interesting questions arising in the study of closed two-sided ideals of the algebra $\mathscr{L}(\mathfrak{B})$ is the question of the relation between the ideal $F(\mathfrak{B})$ of Φ-admissible perturbations and the ideal $S(\mathfrak{B})$ of strictly singular operators. In general, $F(\mathfrak{B}) \neq S(\mathfrak{B})$ [1]. It is proved in [2] that the ideals $F(\mathfrak{B})$ and $S(\mathfrak{B})$ coincide for a subprojective(1) Banach space $\mathfrak{B}$. Examples of subprojective spaces are L_p $(2 \leq p < \infty)$ [4] and the more general spaces Λ_p $(2 \leq p < \infty)$ [5]. At the same time the equality $F(\mathfrak{B}) = S(\mathfrak{B})$ was established also for certain spaces $\mathfrak{B}$ which are not subprojective (in [6] for L_1 and C, and in [7] for L_p, $1 < p < 2$).

Our goal is to obtain the same result for the space $\Lambda_p(\varphi)$ $(1 < p < \infty)$ under certain restrictions on the function $\varphi(t)$. In §1 we derive a theorem on complemented subspaces of $\Lambda_p(\varphi)$ $(1 < p < 2)$. In §2 we establish a theorem on coincidence of the ideals $F(\Lambda_p(\varphi))$ and $S(\Lambda_p(\varphi))$ $(1 < p < 2)$.

§**1.** Let $\varphi(t)$ be an increasing concave function on $[0, 1]$ with $\varphi(0) - \varphi(+0) = 0$. For any $p \in [1, \infty)$ let $\Lambda_p(\varphi)$ denote the Banach space of all measurable functions $x(t)$ on $[0, 1]$ such that

$$\|x\|_{\Lambda_p} = \left(\int_0^1 x^{*^p}(t)\varphi'(t)\,dt\right)^{1/p},$$

where $x^*(t)$ is the nonincreasing function equimeasurable with $|x(t)|$ [8].

1980 *Mathematics Subject Classification* (1985 *Revision*). Primary 46E30, 47D30; Secondary 47A53, 46B15.

Translation of Investigations in Functional Analysis and Differential Equations (B. A. Shcherbakov, editor), "Shtiintsa", Kishinev, 1984, pp. 112–121; MR **86m**: 46029.

(1)A Banach space $\mathfrak{B}$ is said to be *subprojective* [3] if any infinite-dimensional subspace of it contains an infinite-dimensional complemented subspace of $\mathfrak{B}$.

It is easy to see that $\Lambda_p(\varphi) \subset L_p$ and that there exists a constant c such that

$$\|x\|_{L_p} \leq c\|x\|_{\Lambda_p} \qquad (x \in \Lambda_p(\varphi)). \tag{1}$$

If a sequence $\{z_n\}_1^\infty$ converges weakly to z in $\Lambda_p(\varphi)$, then we write $\text{w-lim}_{n\to\infty} z_n = z$.

Denote by $[z_n]_{n\in J}$ ($J \subset N$) the closed linear span of the functions $\{z_n\}_{n\in J}$ in the space $\Lambda_p(\varphi)$.

THEOREM 1. *Suppose that $\Lambda_p(\varphi)$ $(1 < p < 2)$ has an unconditional basis. Then any subspace X isomorphic to l_2 contains an infinite-dimensional complemented subspace of $\Lambda_p(\varphi)$.*

PROOF. Since X does not contain subspaces isomorphic to l_p, it follows from Corollary 4 in [9] and from (1) that the norms of the spaces $\Lambda_p(\varphi)$ and L_p are equivalent on X, i.e.,

$$c_1\|x\|_{\Lambda_p} \leq \|x\|_{L_p} \leq c\|x\|_{\Lambda_p} \qquad (c_1 > 0,\ x \in X). \tag{2}$$

Let $\{e_j\}_1^\infty$ be the standard basis in l_2, and A an isomorphism of l_2 onto X. Then $x_j = Ae_j$ $(j = 1, 2, \dots)$ is a basis for x, and

$$\|A^{-1}\|^{-1} \leq \|x_j\|_{\Lambda_p} \leq \|A\| \qquad (j = 1, 2, \dots). \tag{3}$$

It is known ([10], p. 63) that there exists a constant $0 < c(p) < 1$ such that the modulus of convexity $\delta_{L_p}(\varepsilon)$ $(0 < \varepsilon \leq 2)$ of the space L_p $(1 < p < 2)$ satisfies the inequality

$$\delta_{L_p}(\varepsilon) \geq c^2(p)\varepsilon^2 \qquad (0 < \varepsilon \leq 2). \tag{4}$$

Let $\{h_j\}_1^\infty$ be an unconditional basis for $\Lambda_p(\varphi)$, let K be its basis constant, and let K_1 be its unconditional constant (11, pp. 2, 18). As is known, $K_1 \geq K \geq 1$.

We choose a sequence $\{\eta_j\}_1^\infty$ of numbers such that

$$0 < \eta_j < \min\left(\frac{c_1 c(p)}{2^{j+6}K_1K\|A\|\,\|A^{-1}\|^2 c}, \frac{c_1}{2c\|A^{-1}\|}\right). \tag{5}$$

Since $\Lambda_p(\varphi)$ $(p > 1)$ is a reflexive space [8], $\text{w-lim}_{n\to\infty} x_n = 0$. But then (3) and the proof of Proposition 1.a.12 in [11] give us that there exist subsequences $\{x_{n_k}\}_1^\infty$ and a block basis

$$u_j = \sum_{i=m_{j-1}+1}^{m_j} a_i h_i$$

of the basis $\{h_j\}_1^\infty$ such that

$$\|x_{n_j} - u_j\|_{\Lambda_p} < \eta_j \qquad (j = 1, 2, \dots). \tag{6}$$

It follows from (3), (5), and (6) that

$$\|u_j\|_{\Lambda_p} \geq \|x_{n_j}\|_{\Lambda_p} - \|x_{n_j} - u_j\|_{\Lambda_p} \geq \frac{1}{\|A^{-1}\|} - \frac{1}{2\|A^{-1}\|} = \frac{1}{2\|A^{-1}\|} \qquad (j = 1, 2, \dots).$$

Since

$$\sum_{j=1}^{\infty}\left\|\frac{u_j}{\|u_j\|_{\Lambda_p}}-\frac{x_{n_j}}{\|u_j\|_{\Lambda_p}}\right\|_{\Lambda_p}\le 2\|A^{-1}\|\sum_{j=1}^{\infty}\frac{c_1c(p)}{2^{j+6}K_1K\|A\|\,\|A^{-1}\|^2c}$$
$$=\frac{c_1c(p)}{32K_1K\|A\|\,\|A^{-1}\|c}<\frac{1}{2K},\tag{7}$$

it follows from Theorem 1.a.9 in [11] that $\{x_{n_j}\}_1^{\infty}$ is a basic sequence equivalent to $\{u_j\}_1^{\infty}$. Further, if B is an isomorphism of $[u_j]_1^{\infty}$ onto $[x_{n_j}]_1^{\infty}$, $Bu_j=x_{n_j}$ $(j=1,2,\dots)$, then $\|B\|\le 2$ and $\|B^{-1}\|\le 2$.

If V is an isomorphic operator acting in l_2 such that $Ve_{n_j}=e_j$ $(j=1,2,\dots)$, then $Du_j\equiv VA^{-1}Bu_j=e_j$ $(j=1,2,\dots)$, and also $\|D\|\le 2\|A^{-1}\|$ and $\|D^{-1}\|\le 2\|A\|$. This implies that

$$\frac{1}{2\|A^{-1}\|}\left(\sum_j|\alpha_j|^2\right)^{1/2}\le\left\|\sum_j\alpha_ju_j\right\|_{\Lambda_p}\le 2\|A\|\left(\sum_j|\alpha_j|^2\right)^{1/2}.\tag{8}$$

Let $\{g_j\}_1^{\infty}$ be the sequence biorthogonal to $\{h_j\}_1^{\infty}$, and let

$$P_jx=\sum_{i=m_{j-1}+1}^{m_j}g_i(x)h_i.$$

It is easy to see that the P_j $(j=1,2,\dots)$ project $\Lambda_p(\varphi)$ onto $H_j=[h_i]_{i=m_{j-1}+1}^{m_j}$, and $\|P_j\|\le 2K$ $(j=1,2,\dots)$.

Consider the finite-dimensional space H_j with the norm $\|x\|_{L_p}$. Let v_j denote a functional in H_j^* such that $\|v_j\|=1$, $v_j(u_j)=\|u_j\|_{L_p}$, and let $Q_jx=\frac{v_j(x)}{\|u_j\|_{L_p}}u_j$ $(x\in H_j)$. Obviously, Q_j projects H_j onto the one-dimensional subspace $[u_j]$, and $\|Q_j\|_{L_p}=1$.

We show that $P=\sum_1^{\infty}Q_jP_j$ is the projection projecting $\Lambda_p(\varphi)$ onto $[u_j]_1^{\infty}$ [12].

Let $x\in\Lambda_p(\varphi)$. Then $\sum_1^{\infty}P_jx$ converges unconditionally in $\Lambda_p(\varphi)$. It follows from (1) that this series also converges unconditionally in L_p. We use a refined variant of a theorem of Kadets ([10], pp. 69, 70) given in [13]:

$$\sum_{j=1}^{\infty}\delta_{L_p}\left(\frac{2\|P_jx\|_{L_p}}{\max\limits_{\varepsilon_j=\pm1}\left\|\sum_{j=1}^{\infty}\varepsilon_jP_jx\right\|_{L_p}}\right)\le 1.$$

This and (4) and (1) give us that

$$\sum_{j=1}^{\infty}\|P_jx\|_{L_p}^2 \leq \frac{1}{4c^2(p)}\left(\max_{\varepsilon_j=\pm1}\left\|\sum_{j=1}^{\infty}\varepsilon_jP_jx\right\|_{L_p}\right)^2$$
$$\leq \frac{c^2}{4c^2(p)}\left(\max_{\varepsilon_j=\pm1}\left\|\sum_{j=1}^{\infty}\varepsilon_jP_jx\right\|_{\Lambda_p}\right)^2. \tag{9}$$

It now follows from Proposition 1.c.7 in [11] that

$$\sum_{j=1}^{\infty}\|P_jx\|_{L_p}^2 \leq \frac{c^2}{4c^2(p)}4K_1^2\left\|\sum_{j=1}^{\infty}P_jx\right\|_{\Lambda_p}^2,$$

i.e.,

$$\left(\sum_{j=1}^{\infty}\|P_jx\|_{L_p}^2\right)^{1/2} \leq \frac{K_1c}{c(p)}\|x\|_{\Lambda_p}.$$

Since $Q_jP_jx = \lambda_ju_j$, it follows that $|\lambda_j|\,\|u_j\|_{L_p} \leq \|P_jx\|_{L_p}$. Moreover, by (1) and (2),

$$\|u_j\|_{L_p} \geq \|x_{n_j}\|_{L_p} - \|x_{n_j}-u_j\|_{L_p} \geq c_1\|x_{n_j}\|_{\Lambda_p} - c\|x_{n_j}-u_j\|_{\Lambda_p}$$
$$\geq \frac{c_1}{\|A^{-1}\|} - c\eta_j \geq \frac{c_1}{\|A^{-1}\|} - \frac{c_1}{2\|A^{-1}\|} = \frac{c}{2\|A^{-1}\|}.$$

Therefore,

$$\|P_jx\|_{L_p} \geq \frac{c_1}{2\|A^{-1}\|}|\lambda_j| \qquad (j = 1, 2, \dots),$$

and then

$$\left(\sum_{j=1}^{\infty}|\lambda_j|^2\right)^{1/2} \leq \frac{2\|A^{-1}\|}{c_1}\left(\sum_{j=1}^{\infty}\|P_jx\|_{L_p}^2\right)^{1/2} \leq \frac{2\|A^{-1}\|K_1c}{c_1c(p)}\|x\|_{\Lambda_p}.$$

From this and (8),

$$\left\|\sum_{j=1}^{\infty}\lambda_ju_j\right\|_{\Lambda_p} \leq \frac{4\|A\|\,\|A^{-1}\|K_1c}{c_1c(p)}\|x\|_{\Lambda_p},$$

i.e.,

$$\left\|\sum_{j=1}^{\infty}\lambda_ju_j\right\|_{\Lambda_p} \leq c_2\|x\|_{\Lambda_p},$$

or

$$\|Px\|_{\Lambda_p} \leq c_2\|x\|_{\Lambda_p} \qquad \left(c_2 = \frac{4\|A\|\,\|A^{-1}\|K_1c}{c_1c(p)}\right). \tag{10}$$

Obviously, $P^2 = P$, and hence P is the operator projecting $\Lambda_p(\varphi)$ onto $[u_j]_1^\infty$. By (7) and (10),

$$\sum_{j=1}^{\infty}\left\| \frac{x_{n_j}}{\|u_j\|_{\Lambda_p}} - \frac{u_j}{\|u_j\|_{\Lambda_p}}\right\|_{\Lambda_p} < \frac{1}{8K}\cdot\frac{c_1c(p)}{4K_1c\|A\|\,\|A^{-1}\|} < \frac{1}{8K\|P\|},$$

and the basis constant of $\{u_j\}_1^\infty$ does not exceed K; therefore, this and Theorem 1.a.9 in [11] give us that $[x_{n_j}]_1^\infty$ is complemented in $\Lambda_p(\varphi)$.

The theorem is proved.

An analogous theorem was obtained in [12] for $\varphi(t) = t$. Various results on complemented subspaces of $\Lambda_p(\varphi)$ were established in [5] and [9].

If we introduce in $\Lambda_p(\varphi)$ the dilation operator σ_τ $(0 < \tau < \infty)$ by the equalities

$$(\sigma_\tau x)(t) = x(t\tau^{-1}) \quad (\tau \geq 1),$$

$$(\sigma_\tau x)(t) = \begin{cases} x(t\tau^{-1}), & 0 < t \leq \tau \\ 0, & \tau < t \leq 1 \end{cases} \quad (0 < \tau < 1),$$

then, as in [14] (Chapter II, §4.3), we see that

$$\|\sigma_\tau\| = \begin{cases} \left(\sup\limits_{0<t\leq 1} \dfrac{\varphi(\tau t)}{\varphi(t)}\right)^{1/p} & (0 < \tau \leq 1), \\ \left(\sup\limits_{0<t\leq \tau^{-1}} \dfrac{\varphi(\tau t)}{\varphi(t)}\right)^{1/p} & (\tau > 1). \end{cases} \tag{11}$$

Since $\varphi(t)$ is a concave function on $[0, 1]$, it follows that $\varphi(\tau t) \leq \tau\varphi(t)$, $(\tau > 1)$, and thus $\|\sigma_\tau\| = \tau^{1/p}$ $(\tau > 1)$. Hence,

$$\|\sigma_\tau\| = o(\tau) \quad \text{as} \quad \tau \to \infty. \tag{12}$$

It follows from (11), (12), and [14] (Chapter II, §9.4) that the spaces $\Lambda_p(\varphi)$ $(1 < p < \infty)$ have an unconditional basis (the Haar basis) if and only if

$$\lim_{\tau\to 0}\sup_{0<\tau\leq 1} \frac{\varphi(\tau t)}{\varphi(t)} = 0. \tag{13}$$

Therefore, Theorem 1 can also be formulated as follows.

THEOREM 1′. *Suppose that* $1 < p < 2$, *and assume condition* (13). *Then any subspace* X *of* $\Lambda_p(\varphi)$ *isomorphic to* l_2 *contains an infinite-dimensional complemented subspace of* $\Lambda_p(\varphi)$.

§2. Let $\mathscr{L}(\mathfrak{B})$ be the Banach algebra of all bounded linear operators acting in the Banach space $\mathfrak{B}$.

An operator $T \in \mathscr{L}(\mathfrak{B})$ is said to be *strictly singular* if for any infinite-dimensional subspace $E \subset \mathfrak{B}$ the restriction $T|E$ of T to E is not an isomorphism. The set of all such operators is denoted by $S(\mathfrak{B})$.

For any $T \in \mathscr{L}(\mathfrak{B})$ let $\operatorname{Im} T$ denote the range of T, and $\ker T$ the kernel of T. An operator $T \in \mathscr{L}(\mathfrak{B})$ is called a Φ-*operator* if $\operatorname{Im} T = \overline{\operatorname{Im} T}$, $\dim\ker T < \infty$, and $\dim \mathfrak{B}|_{\operatorname{Im} T} < \infty$. If $\mathfrak{M}$ is a subspace of $\mathfrak{B}$, then let $T(\mathfrak{M}) = \operatorname{Im} T|\mathfrak{M}$.

Let $F(\mathfrak{B})$ be the set of all Φ-admissible perturbations, i.e., $T \in F(\mathfrak{B})$ if for any Φ-operator $A \in \mathscr{L}(\mathfrak{B})$ the operator $A + T$ is also a Φ-operator.

The sets $F(\mathfrak{B})$ and $S(\mathfrak{B})$ are closed two-sided ideals of the algebra $\mathscr{L}(\mathfrak{B})$, and $S(\mathfrak{B}) \subset \mathscr{L}(\mathfrak{B})$ (see [1] and [15]).

LEMMA 1. *Suppose that $\mathfrak{B}$ is a Banach space, $T \in L(\mathfrak{B})$, $\mathfrak{M}$ is an infinite-dimensional subspace of $\mathfrak{B}$, and $T_1 = T|\mathfrak{M}$ is an isomorphism. If $\operatorname{Im} T_1$ contains an infinite-dimensional complemented subspace X of $\mathfrak{B}$, then $T \notin F(\mathfrak{B})$.*

PROOF. Suppose that P is a projection onto X, and let $\widetilde{T} = T_1^{-1}P$. Since $(I - T\widetilde{T})x = 0$ $(x \in X)$, it follows that $X \subset \ker(I - T\widetilde{T})$, and hence $\dim\ker(I - T\widetilde{T}) = \infty$. This implies that $T \notin F(\mathfrak{B})$, for otherwise $T\widetilde{T} \in F(\mathfrak{B})$, and $I - T\widetilde{T}$ must be a Φ-operator. The lemma is proved.

The main result in this paper is the following Theorem 2. Its proof uses some methods of Weis from [7]. We remark that a result of Weis can be obtained from Theorem 2 for $\varphi(t) = t$.

THEOREM 2. *Suppose that $\Lambda_p(\varphi)$ $(1 < p < 2)$ has an unconditional basis and $L_2 \subset \Lambda_p(\varphi)$. Then $F(\Lambda_p(\varphi)) = S(\Lambda_p(\varphi))$.*

PROOF. Obviously, it suffices to prove that $F(\Lambda_p(\varphi)) \subset S(\Lambda_p(\varphi))$. We consider an operator $T \in F(\Lambda_p(\varphi))$ and assume that $T \notin S(\Lambda_p(\varphi))$. Then there exists an infinite-dimensional subspace $\mathfrak{M}$ such that $T|\mathfrak{M}$ is an isomorphism.

We consider two cases:

1) $\mathfrak{M}$ contains a subspace isomorphic to l_p. Then $T(\mathfrak{M})$ also contains a subspace isomorphic to l_p. By [9], $T(\mathfrak{M})$ contains a complemented subspace isomorphic to l_p. From Lemma 1, $T \notin F(\Lambda_p(\varphi))$, which is impossible.

2) $\mathfrak{M}$ does not contain subspaces isomorphic to l_p.

Since $\Lambda_p(\varphi)$ has an unconditional basis, $\mathfrak{M}$ contains an unconditional basic sequence $\{x_n\}_1^\infty$ $(\|x_n\|_{\Lambda_p} = 1,\ n = 1, 2, \dots)$ in view of Proposition 1.a.11 in [11].

It follows from [9] (Theorem 1 and Corollary 2) that if $\mathfrak{M} \subset \Lambda_p(\varphi)$ $(1 < p < 2)$ does not contain subspaces isomorphic to l_p, then

$$\lim_{\delta\to 0} \sup_{\substack{x\in\mathfrak{M}\\ \|x\|_{\Lambda_p}=1}} \|x^*\chi_{[0,\delta]}\|_{\Lambda_p} = 0,$$

where $\chi_e(t)$ is the characteristic function of a measurable set $e \subset [0, 1]$, and $x^*(t)$ is the nonincreasing function equimeasurable with $|x(t)|$. This implies that for any $\varepsilon > 0$ there exists a $\delta = \delta(\varepsilon) > 0$ such that

$$\|x_n^*\chi_{[0,\delta]}\|_{\Lambda_p} < \varepsilon \qquad (n = 1, 2, \dots). \tag{14}$$

Let $A_n = \{t \in [0, 1] : |x_n(t)| > M\}$ and $a_n = \operatorname{meas} A_n$ $(n = 1, 2, \dots)$, where $\operatorname{meas} e$ is the Lebesgue measure of a set $e \subset [0, 1]$. Since

$$1 = \|x_n\|_{\Lambda_p}^p = \int_0^1 x_n^{*^p}(t)\varphi'(t)\,dt > \int_0^{a_n} x_n^{*^p}(t)\varphi'(t)\,dt \geq M^p\varphi(a_n),$$

it follows that $\varphi(a_n) \le M^{-p}$ $(n = 1, 2, \dots)$. Consequently, for sufficiently large $M > 0$,

$$a_n < \delta \qquad (n = 1, 2, \dots). \tag{15}$$

Let

$$\bar{x}_n(t) = \begin{cases} x_n(t), & \text{if } |x_n(t)| \le M; \\ 0, & \text{if } |x_n(t)| > M. \end{cases}$$

Since $\|\bar{x}_n\|_{\Lambda_p} \le \|x_n\|_{\Lambda_p} = 1$ and $\Lambda_p(\varphi)$ $(p > 1)$ are reflexive spaces, it follows that $\{\bar{x}_n\}_1^\infty$ contains a weakly convergent subsequence. It can be assumed without loss of generality that $\text{w-lim}_{n\to\infty} \bar{x}_n = x$.

Since $\{x_n\}_1^\infty$ is a basic sequence in a reflexive space, it follows that $\text{w-lim}_{n\to\infty} x_n = 0$, and hence

$$\operatorname*{w-lim}_{n\to\infty}(\bar{x}_n - x_n) = 0.$$

By (14) and (15),

$$\|x_n - \bar{x}_n\|_{\Lambda_p} = \|x_n \chi_{A_n}\|_{\Lambda_p} = \|x_n^* \chi_{[0, a_n]}\|_{\Lambda_p} < \varepsilon \quad (n = 1, 2, \dots),$$

and then $\|x\|_{\Lambda_p} \le \varepsilon$.

We choose a set $B \subset [0, 1]$ such that $\operatorname{meas} B > 1 - \delta$ and $x(t)$ is bounded on B ($|x(t)| \le M_0$, $t \in B$). Let $y_n = (\bar{x}_n - x_n)\chi_B$. Then

$$|y_n(t)| \le M + M_0. \tag{16}$$

Since $y_n = (\bar{x}_n - x_n)\chi_B + x_n - x\chi_B - x_n \chi_{[0,1]\setminus B}$ and $\operatorname{meas}([0, 1]\setminus B) < \delta$, it follows from (14) that

$$\|x_n \chi_{[0,1]\setminus B}\|_{\Lambda_p} \le \|x_n^* \chi_{[0,\delta]}\|_{\Lambda_p} < \varepsilon.$$

Moreover, it is clear that

$$\|(\bar{x}_n - x_n)\chi_B\|_{\Lambda_p} \le \|\bar{x}_n - x_n\|_{\Lambda_p} < \varepsilon, \qquad \|x\chi_B\|_{\Lambda_p} \le \|x\|_{\Lambda_p} \le \varepsilon,$$

and therefore

$$\|y_n\|_{\Lambda_p} \ge \|x_n\|_{\Lambda_p} - \|(\bar{x}_n - x_n)\chi_B\|_{\Lambda_p} - \|x\chi_B\|_{\Lambda_p} - \|x_n \chi_{[0,1]\setminus B}\|_{\Lambda_p} \ge 1 - 3\varepsilon. \tag{17}$$

Also,

$$\|y_n - x_n\|_{\Lambda_p} \le \|(\bar{x}_n - x_n)\chi_B\|_{\Lambda_p} + \|x\chi_B\|_{\Lambda_p} + \|x_n \chi_{[0,1]\setminus B}\|_{\Lambda_p} < 3\varepsilon. \tag{18}$$

Since the $\Lambda_p(\varphi)$ $(p > 1)$ are reflexive and separable, it follows that $(\Lambda_p)^*$ is separable and, moreover, $(\Lambda_p(\varphi))^*$ is isometric to the associated space $(\Lambda_p)'$ ([14], and also [10], p. 29).

Thus, if $\tilde{f} \in (\Lambda_p(\varphi))^*$, then there exists an $f \in (\Lambda_p)'$ such that

$$\tilde{f}(x) = \int_0^1 x(t) f(t)\, dt.$$

Since $\text{w-lim}_{n\to\infty}(\bar{x}_n - x) = 0$ and $(\Lambda_p)'$ is a symmetric space [14], it follows that $\text{w-lim}_{n\to\infty} y_n = 0$, and then

$$\operatorname*{w-lim}_{n\to\infty} T y_n = 0. \tag{19}$$

By assumption, $T|\mathfrak{M}$ is an isomorphism, and so there exists a $c_3 > 0$ such that $\|Tx_n\|_{\Lambda_p} \geq c_3 > 0$. From this and (18),

$$\|Ty_n\|_{\Lambda_p} \geq \|Tx_n\|_{\Lambda_p} - \|T(y_n - x_n)\|_{\Lambda_p} \geq c_3 - 3\varepsilon\|T\| \geq c_4 > 0 \tag{20}$$

for sufficiently small $\varepsilon > 0$.

It follows from (19), (20), and Proposition 1.a.12 in [11] that some subsequence of $\{Ty_n\}_1^\infty$ is equivalent to a block basis of the unconditional basis $\{h_n\}_1^\infty$ for the space $\Lambda_p(\varphi)$. We assume that this is true of the sequence $\{Ty_n\}_1^\infty$ itself.

Since w-$\lim_{n\to\infty} y_n = 0$ and $L_\infty \subset (\Lambda_p)'$, it follows that

$$\lim_{n\to\infty} \int_0^1 y_n(t) f(t)\, dt = 0 \qquad (f \in L_\infty).$$

Moreover,

$$\|y_n\|_{L_2} \leq \sup_t |y_n(t)| \leq M + M_0$$

by (16). Since L_∞ is dense in L_2, the sequence $\{y_n\}_1^\infty$ converges weakly to zero in L_2.

By a condition of the theorem, $L_2 \subset \Lambda_p(\varphi)$. Hence, L_2 is continuously imbedded in $\Lambda_p(\varphi)$ [14], i.e., there exists a constant $c_5 > 0$ such that

$$\|z\|_{L_2} \geq c_5 \|z\|_{\Lambda_p} \qquad (z \in L_2), \tag{21}$$

and $\|y_n\|_{L_2} \geq (1 - 3\varepsilon)c_5$ in view of (17).

It now follows from Proposition 1.a.12 in [11] that some subsequence of $\{y_n\}_1^\infty \subset \Lambda_p(\varphi)$ is equivalent to the standard basis in l_2. It can be assumed without loss of generality that the sequence $\{y_n\}_1^\infty$ itself has this property. Consequently, there exist constants $c_7 \geq c_6 > 0$ such that

$$c_6 \left(\sum_{j=1}^m |\alpha_j|^2 \right)^{1/2} \leq \left\| \sum_{j=1}^m \alpha_j y_j \right\|_{\Lambda_p} \leq c_7 \left(\sum_{j=1}^m |\alpha_j|^2 \right)^{1/2}. \tag{22}$$

From this and (21) we get

$$\left\| \sum_{j=1}^m \alpha_j T y_j \right\|_{\Lambda_p} \leq \|T\| \left\| \sum_{j=1}^m \alpha_j y_j \right\|_{\Lambda_p} \leq \frac{\|T\|}{c_5} \left\| \sum_{j=1}^m \alpha_j y_j \right\|_{L_2} \leq c_8 \left(\sum_{j=1}^m |\alpha_j|^2 \right)^{1/2}. \tag{23}$$

As in the proof of (9), we have that

$$\begin{aligned} \sum_{j=1}^m |\alpha_j|^2 \|Ty_j\|_{L_p}^2 &\leq \frac{c^2}{4c^2(p)} \left(\max_{\varepsilon_j = \pm 1} \left\| \sum_{j=1}^m \varepsilon_j \alpha_j, y_j \right\|_{\Lambda_p} \right)^2 \\ &\leq \frac{c^2}{4c^2(p)} K_2^2 \left\| \sum_{j=1}^m \alpha_j T y_j \right\|_{\Lambda_p}^2, \end{aligned}$$

where K_2 is the unconditional constant of $\{Ty_j\}_1^\infty$. Consequently,

$$\left\|\sum_{j=1}^{m}\alpha_j Ty_j\right\|_{\Lambda_p} \geq c_9\left(\sum_{j=1}^{m}|\alpha_j|^2\|Ty_j\|_{L_p}^2\right)^{1/2}. \tag{24}$$

We get a lower estimate for $\|Ty_j\|_{L_p}$. Since $T(\mathfrak{M})$ (just like $\mathfrak{M}$) does not contain subspaces isomorphic to l_p, it follows from [9] (Corollary 2 and Lemma 1) that there exists a $c_{10} > 0$ such that

$$\|z\|_{L_p} \geq c_{10}\|z\|_{\Lambda_p} \qquad (z \in T(\mathfrak{M})).$$

From this and (18) and (1) we get

$$\|Tx_n\|_{L_p} \geq c_{10}\|Tx_n\|_{\Lambda_p} \geq c_{10}c_3 = c_{11},$$

and

$$\begin{aligned}\|Ty_n\|_{L_p} &\geq \|Tx_n\|_{L_p} - \|Ty_n - Tx_n\|_{L_p} \geq c_{11} - c\|Ty_n - Tx_n\|_{\Lambda_p} \\ &\geq c_{11} - c\|T\|\,\|y_n - x_n\|_{\Lambda_p} \geq c_{11} - 3\varepsilon c\|T\| \geq c_{12} > 0\end{aligned}$$

for sufficiently small $\varepsilon > 0$.

Taking account of the last inequality in (24), we get that

$$\left\|\sum_{j=1}^{m}\alpha_j Ty_j\right\|_{\Lambda_p} \geq c_9c_{12}\left(\sum_{j=1}^{m}|\alpha_j|^2\right)^{1/2}. \tag{25}$$

Let $\widetilde{\mathfrak{M}} = [y_j]_1^\infty$. It follows from (22), (23), and (25) that $T|\widetilde{\mathfrak{M}}$ is an isomorphism, and, moreover, $T(\widetilde{\mathfrak{M}})$ is isomorphic to l_2. By Theorem 1, $T(\widetilde{\mathfrak{M}})$ contains an infinite-dimensional complemented subspace. But then $T \notin F(\Lambda_p(\varphi))$ by Lemma 1, and the choice of T is contradicted.

The theorem is proved.

It is easy to determine a condition sufficient for $L_2 \subset \Lambda_p(\varphi)$ $(1 < p < 2)$. For example, this is so if $\varphi' \in L_2[0, 1]$. Therefore, we have

THEOREM 3. *Suppose that condition* (12) *holds,* $p \in (1, 2)$, *and* $\varphi' \in L_{2/(2-p)}[0, 1]$. *Then* $F(\Lambda_p(\varphi)) = S(\Lambda_p(\varphi))$.

BIBLIOGRAPHY

1. S. R. Caradus, W. E. Pfaffenberger, and Bertram Yood, *Calkin algebras and algebras of operators on Banach spaces*, Lecture Notes Pure Appl. Math., vol. 9, Marcel Dekker, New York, 1974.

2. William Pfaffenberger, *On the ideals of strictly singular and inessential operators*, Proc. Amer. Math. Soc. **25** (1970), 603–607.

3. R. J. Whitley, *Strictly singular operators and their conjugates*, Trans. Amer. Math. Soc. **113** (1964), 252–261.

4. M. I. Kadec [Kadets] and A. Pełczyński, *Bases, lacunary sequences and complemented subspaces in the spaces L_p*, Studia Math. **21** (1961/62), 161–176.

5. T. Figiel, W. B. Johnson, and L. Tzafriri, *On Banach lattices and spaces having local unconditional structure, with applications to Lorentz function spaces*, J. Approximation Theory **13** (1975), 395–412.

6. V. D. Mil'man, *Operators of class* C_0 *and* C_0^*, Teor. Funktsiĭ, Funktsional. Anal. i Prilozhen. Vyp. 10 (1970), 15–26. (Russian)

7. L. Weis, *On perturbations of Fredholm operators in* $L_p(\mu)$*-spaces*, Proc. Amer. Math. Soc. **67** (1977), 287–292.

8. G. G. Lorentz, *On the theory of spaces* Λ, Pacific J. Math. **1** (1951), 411–429.

9. S. Ya. Novikov, E. M. Semenov, and E. V. Tokarev, *The structure of subspaces of the space* $\Lambda_p(\varphi)$, Dokl. Akad. Nauk SSSR **247** (1979), 552–554; English transl. in Soviet Math. Dokl. **20** (1979).

10. Joram Lindenstrauss and Lior Tzafriri, *Classical Banach spaces.* II, Springer-Verlag, 1979.

11. ——, *Classical Banach spaces.* I, Springer-Verlag, 1977.

12. A. Pełczyński and H. P. Rosenthal, *Localization techniques in* L^p *spaces*, Studia Math. **52** (1974/75), 263–289.

13. V. I. Gurariĭ and A. S. Markus, *Some remarks about vector and operator series in uniformly convex and uniformly smooth spaces*, Mat. Issled. **10** (1975), vyp. 1 (35), 126–134. (Russian)

14. S. G. Kreĭn, Yu. I. Petunin, and E. M. Semenov, *Interpolation of linear operators*, "Nauka", Moscow, 1978; English transl., Amer. Math. Soc., Providence, R.I., 1982.

15. Albrecht Pietsch, *Operator ideals*, VEB Deutscher Verlag Wiss., Berlin, 1978.

Translated by H. H. McFADEN

Amer. Math. Soc. Transl.
(2) Vol. **142**, 1989

A Topology on the Set of Maximal Ideals of a Banach PI-Algebra

M. B. ABALOVICH AND N. YA. KRUPNIK

§**1.** Let $\mathfrak{A}$ be a Banach algebra with identity e over the field $\mathbb{C}$ of complex numbers, and denote by $\mathscr{R}(\mathfrak{A})$, $G\mathfrak{A}$, and $\mathfrak{M}$ its radical, group of invertible elements, and the set of maximal two-sided ideals of the quotient algebra $\widetilde{\mathfrak{A}} = \mathfrak{A}/\mathscr{R}(\mathfrak{A})$, respectively. Let $L(\mathfrak{B})$ denote the algebra of all bounded linear operators acting in a Banach space $\mathfrak{B}$. A collection $\{\nu_\tau\}$ $(\tau \in \mathscr{T})$ of representations of the algebra $\mathfrak{A}$ $(\nu_\tau : \mathfrak{A} \to L(\mathfrak{B}_\tau))$ is said to be *sufficient* if it is responsible for the invertibility of elements in the algebra $\mathfrak{A}$: $x \in G\mathfrak{A} \Leftrightarrow (\nu_\tau(x) \in GL(\mathfrak{B}_\tau)\ \forall \tau \in \mathscr{T})$. If $\mathfrak{A}$ has a sufficient collection $\{\nu_\tau\}$ $(\tau \in \mathscr{T})$ of representations and $\dim \mathfrak{B}_\tau \leq n\ \forall \tau \in \mathscr{T}$, then we write $\mathfrak{A} \in \pi_n$. It has been established in [1] and [2] that $\mathfrak{A} \in \pi_n$ for some n if and only if $\widetilde{\mathfrak{A}}$ is an algebra with a polynomial identity (a PI-algebra [3]). If this condition holds, then for each $M \in \mathfrak{M}$ the quotient algebra $\widetilde{\mathfrak{A}}/M$ is isomorphic to $L(\mathfrak{B}_M)$, where $\dim \mathfrak{B}_M \subseteq [d/2]$, and d is the degree of the polynomial giving rise to the identity on $\widetilde{\mathfrak{A}}$. The collection of natural homomorphisms $\widetilde{\mathfrak{A}} \to \widetilde{\mathfrak{A}}/M$ induces a sufficient collection of irreducible representations of $\mathfrak{A}$. Choosing some basis in each space $\mathfrak{B}_M$, we can construct a sufficient collection of matrix representations of the algebra $\mathfrak{A}$: $\nu_M(x) = (\nu_{ik}(M, x))_{i,k=1}^{l}$, where $l = l(M) = \dim \mathfrak{B}_M$.

Banach PI-algebras $(\widetilde{\mathfrak{A}})$ are a natural generalization of commutative Banach algebras.

In this article we investigate a topology on the set $\mathfrak{M}$ of maximal two-sided ideals of a Banach PI-algebra $(\widetilde{\mathfrak{A}})$, and the Gel'fand representation of the algebra $\mathfrak{A}$ by the algebra of continuous matrix-valued functions on $\mathfrak{M}$. We obtain a matrix analogue of the well-known Gel'fand–Naĭmark theorem on

1980 *Mathematics Subject Classification* (1985 *Revision*). Primary 46H10, 46H15, 46J20.
Translation of Investigations in Functional Analysis and Differential Equations (B. A. Shcherbakov, editor), "Shtiintsă, Kishinev, 1984, pp. 3–11; MR **87e**:46063.

the isometry between a commutative Banach C^*-algebra and the algebra of all continuous functions on the compact space of maximal ideals.

§**2.** We consider some properties of matrix representations of $\mathfrak{A}$.

LEMMA 1. *An irreducible matrix representation of the algebra $\mathfrak{A}$ is continuous.*

PROOF. Let ν be an irreducible representation, $\nu(x) = (\alpha_{ik})_{i,k=1}^n$. By the Burnside theorem, the range of ν is the full matrix algebra $\mathbb{C}^{n\times n}$. Let $e_{jk} \in \mathbb{C}^{n\times n}$ $(j, k = 1, \dots, n)$ be the matrix units, and choose $z_{jk} \in \mathfrak{A}$ such that $\nu(z_{jk}) = e_{jk}$. It can be verified directly that

$$e_{1j}\nu(x)e_{k1} = \begin{pmatrix} \alpha_{jk} & 0 & \cdots & 0 \\ 0 & 0 & \cdots & 0 \\ \vdots & \vdots & & \vdots \\ 0 & 0 & & 0 \end{pmatrix}.$$

We show that

$$|\alpha_{jk}| \leq \|z_{1j}\| \|z_{k1}\| \|x\|. \tag{1}$$

If this is not so, then $|\alpha_{jk}| > \|z_{1j}xz_{k1}\|$; hence, $\|\frac{z_{1j}xz_{k1}}{\alpha_{jk}}\| < 1$. Then $\alpha_{jk}e - z_{1j}xz_{k1} \in G\mathfrak{A}$, and so $\nu(\alpha_{jk}e - z_{1j}xz_{k1}) \in G\mathbb{C}^{n\times n}$, a contradiction. Consequently, (1) holds.

Let $\nu_M(x) = (x_{jk}(M))_1^n$. According to Lemma 1, $|x_{jk}(M)| \leq c\|x\|$, where $c = c(M)$. In what follows, we require "uniform boundedness" of the representations in a sufficient collection. However, this is not always the case (see the corresponding example in [4]). We show that we can choose an equivalent sufficient collection with the required property (with constant $c = 1$). For this we need some auxiliary statements.

LEMMA 2. *Suppose that E is an n-dimensional complex linear space. There exists a basis $e_1, \dots, e_n$ in E such that $|a_{jk}| \leq \|A\|$ $\forall A \in L(E)$, where $(a_{jk})_{j,k=1}^n$ is the matrix of the operator A in this basis.*

PROOF. Choose an Auerbach basis in E, i.e., a basis $\{e_i\}_1^n$ such that $\|e_i\| = \|f_j\| = 1$, where $f_j \in E^*$ and $f_j(e_i) = \delta_{ij}$ ([5], p. 16). It is clear that $a_{ij} = f_i(Ae_j)$; but then

$$|a_{ij}| = |f_i(Ae_j)| \leq \|f_i\| \|A\| \|e_j\| = \|A\|.$$

LEMMA 3[1]. *Suppose that the algebraic norm $\|\cdot\|$ is introduced on $\mathbb{C}^{n\times n}$. Then there is an operator norm $\|\cdot\|_1$ on $C^{n\times n}$ such that $\|A\|_1 \leq \|A\|$ $\forall A \in \mathbb{C}^{n\times n}$.*

PROOF. We regard $A = \|a_{jk}\|_{j,k=1}^n$ as the matrix of some operator in the space of n-columns $\mathbb{C}^n$ with the norm $\|x\|_2 = \|B_x\|$, where we set

$$B_x = \begin{pmatrix} x_1 & \\ \vdots & 0 \\ x_n & \end{pmatrix}$$

[1] The authors are indebted to A. S. Markus for this assertion.

for $x = (x_1, \dots, x_n)^{\mathrm{T}}$. Then

$$\|A\|_1 = \sup_{x\neq 0} \frac{\|Ax\|_2}{\|x\|_2} = \sup_{x\neq 0} \frac{\|(a_{jk})(x_j)\|_2}{\|(x_j)\|_2}$$

$$= \sup_{x\neq 0} \frac{\|(\sum_{k=1}^n a_{jk}x_k)_{j=1}^n\|_2}{\|(x_j)\|_2} = \sup_{x\neq 0} \frac{\left\|\begin{pmatrix} \sum a_{1k}x_k & \\ \vdots & 0 \\ \sum a_{nk}x_k & \end{pmatrix}\right\|}{\left\|\begin{pmatrix} x_1 & \\ \vdots & 0 \\ x_n & \end{pmatrix}\right\|}$$

$$= \sup \frac{\left\|\begin{pmatrix} a_{11}\cdots a_{1n} \\ \cdots\cdots \\ a_{n1}\cdots a_{nn} \end{pmatrix}\begin{pmatrix} x_1 & \\ \vdots & 0 \\ x_n & \end{pmatrix}\right\|}{\left\|\begin{pmatrix} x_1 & \\ \vdots & 0 \\ x_n & \end{pmatrix}\right\|} \leq \sup \frac{\left\|\begin{pmatrix} a_{11}\cdots a_{1n} \\ \cdots\cdots \\ a_{n1}\cdots a_{nn} \end{pmatrix}\right\| \left\|\begin{pmatrix} x_1 & \\ \vdots & 0 \\ x_n & \end{pmatrix}\right\|}{\left\|\begin{pmatrix} x_1 & \\ \vdots & 0 \\ x_n & \end{pmatrix}\right\|}$$

$$= \|A\|.$$

Lemma 4. *Let ν be an irreducible matrix representation of a Banach algebra $\mathfrak{A}$. There exists an equivalent representation ν' such that*

$$|\nu'_{jk}(x)| \leq \|x\| \quad \forall x \in \mathfrak{A}.$$

Proof. Let us introduce in $\mathbb{C}^{n\times n}$ the (algebraic) norm

$$\|\nu(x)\| = \inf_{y\in \operatorname{Ker}\nu} \|x+y\|$$

(here the irreducibility of ν is used). We regard $\nu(x)$ as an operator on $\mathbb{C}^n$ with the standard basis and the norm in Lemma 3 (the same notation is kept for the matrix of the given operator). According to Lemma 3, we define an operator norm $\|\cdot\|_1$ such that $\|\nu(x)\|_1 \leq \|\nu(x)\|$. We introduce the basis $e_1, \dots, e_n$ from Lemma 2 in $\mathbb{C}^n$ and let T be the matrix of transition to this basis. Let $\nu'(x) = T\nu(x)T^{-1}$. Then ν' is the matrix of the operator $\nu(x)$ in the basis $\{e_i\}_1^n$. According to Lemma 2, $|\nu'_{jk}(x)| \leq \|\nu(x)\|_1 \leq \|\nu(x)\| \leq \|x\|$, which is what was required to be proved.

Remark. An analogous but weaker result (with the constant 2^n) was obtained in [6].

Theorem 1. *Suppose that $\mathfrak{A} \in \pi_n$, and let $\{\nu_\tau\}$ $(\tau \in \mathscr{T})$ be a sufficient set for it. Then there exists an equivalent sufficient set $\{\nu'_\tau\}$ $(\tau \in \mathscr{T})$ such that $|x'_{jk}(\tau)| \leq \|x\|$.*

The proof follows directly from Lemma 4.

Let $\mathfrak{A} \in \pi_n$. For each k with $1 \leq k \leq n$ let $\mathfrak{M}_k = \{M \in \mathfrak{M}: l(M) = k\}$. For each $\varepsilon > 0$, $x_1, \dots, x_r \in \mathfrak{A}$, and $M_0 \in \mathfrak{M}$ we consider the sets

$$u_{x_1,\dots,x_r,\varepsilon}(M_0) = \{M \in \mathfrak{M}_k : |x^s_{pq}(M_0) - x^s_{pq}(M)| < \varepsilon; p, q = 1, \dots, k; s = 1, \dots, r\}.$$

It can be verified directly that these sets can serve as a local base of neighborhoods of the point M_0 in a certain topology. This topology is called the Gel'fand topology on $\mathfrak{M}$, and in it $\mathfrak{M}$ becomes a Hausdorff topological space. It is clear that for each k with $1 \leq k \leq n$ the set $\mathfrak{M}_k$ is open and closed in $\mathfrak{M}$. In general, $\mathfrak{M}$ is not compact (an example is given below).

We consider a special case when $\mathfrak{M}$ is compact. We say that $\mathfrak{A}$ *satisfies condition* (α) if $l(M) = n\ \forall M \in \mathfrak{M}$, and $\mathfrak{M}$ is closed in $\mathfrak{N}$ (see the proof of Theorem 2).

THEOREM 2. *Suppose that $\mathfrak{A}$ is a PI-algebra satisfying condition (α). Then $\mathfrak{M}$ is compact (in the Gel'fand topology).*

PROOF. For each ν in $\{\nu_M\}$ we have $|x_{jk}(M)| \leq \|x\|$ by Theorem 1 (otherwise we pass to an equivalent sufficient set). With each $x \in \mathfrak{A}$ we associate the n^2-dimensional disk $\mathscr{D}_x^n = \prod_1^n \mathscr{D}_x$, where $\mathscr{D}_x = \{\lambda \in \mathbb{C}\colon |\lambda| \leq \|x\|\}$. The $\mathscr{D}_x^n$ are compact; consequently, their product $\mathfrak{N} = \prod_{x\in\mathfrak{A}} \mathscr{D}_x^n$ is also compact by Tychonoff's theorem. With each $M \in \mathfrak{M}$ we associate the point $\{\nu_M(x)\}_{x\in\mathfrak{A}}$ in $\mathfrak{N}$. It is clear from a comparison of topologies that this is a homeomorphic imbedding of $\mathfrak{M}$ in $\mathfrak{N}$. Since $\mathfrak{M}$ is closed in $\mathfrak{N}$, the theorem is proved.

Suppose that a net M_α converges, and let $\nu(x) = \lim_\alpha \nu_{M_\alpha}(x)$; then

$$\nu(xy) = \lim_\alpha \nu_{M_\alpha}(xy) = \lim \nu_{M_\alpha}(x) \lim \nu_{M_\alpha}(y) = \nu(x)\nu(y).$$

It is verified similarly that $\nu(x+y) = \nu(x) + \nu(y)$, and $\nu(e) = e$. Let $\mathscr{J} = \operatorname{Ker} \nu$; then $\mathscr{J}$ is a maximal ideal. Indeed, if not, then by passing to the quotient of $\operatorname{Im} \nu$ by the radical and then decomposing the quotient algebra into a direct sum of full matrix algebras, we would get that $\mathfrak{A}/K \cong \mathbb{C}^{r\times r}$ $(r < n)$ for some $K \in \mathfrak{M}$, and this contradicts condition (α). However, $\nu_{\mathscr{J}} \neq \nu$ in general.

§**3.** A functional topology on the set $\hat{\mathfrak{A}}$ of all (equivalence classes of) finite-dimensional irreducible representations of an algebra $\mathfrak{A}$ was introduced and studied in [6]. We give some results from that paper.

For $T \in \hat{\mathfrak{A}}$ let $\dim(T)$ denote the dimension of the representation space E_T. Let

$$\hat{\mathfrak{A}}^{(n)} = \{T \in \hat{\mathfrak{A}}\colon \dim(T) \leq n\};$$
$$\hat{\mathfrak{A}}_n = \{T \in \hat{\mathfrak{A}}\colon \dim(T) = n\}.$$

For each $T \in \hat{\mathfrak{A}}$ let

$$\Phi(T) = \{\varphi \in \mathfrak{A}^*\colon \operatorname{Ker} T \subset \operatorname{Ker} \varphi\}.$$

For example, $\chi_T \in \Phi(T)\ \forall T \in \hat{\mathfrak{A}}$, where χ_T is the character of the representation T, computed by the formula $\chi_T(a) = \operatorname{Tr} T(a)$. A representation $T \in \hat{\mathfrak{A}}$ is uniquely determined by specifying a nonzero functional in $\varphi(T)$ ([6], Proposition 2). The functional topology is determined by a closure operation: If $T \in \hat{\mathfrak{A}}$ and $S \subset \hat{\mathfrak{A}}$, then $T \in \overline{S}$ if $\varphi(T) \subset \overline{(\bigcup_{s\in S} \varphi(s))}$, where the bar denotes the closure in the weak* topology on $\mathfrak{A}^*$.

The set $\hat{\mathfrak{A}}$ has been turned into a topological T_1-space. In general, $\hat{\mathfrak{A}}$ is not Hausdorff; an example is given below. A "measure of non-Hausdorffness" is defined as follows: if a net $\{T^{(i)}\}$ of elements in $\hat{\mathfrak{A}}^{(n)}$ converges to different elements $V_1, \dots, V_p$, then $\sum_{s=1}^{p} \dim(V_s) \leq n$ ([6], Proposition 13).

A *character* on $\mathfrak{A}$ is defined to be a linear functional on $\mathfrak{A}$ of the form $\psi = \sum_{i=1}^{r} n_i \chi_{T_i}$, where the n_i are nonnegative integers, $T_1, \dots, T_r$ are pairwise nonequivalent elements of $\hat{\mathfrak{A}}$, and $\chi_{T_1}, \dots, \chi_{T_r}$ are their characters, defined above. If ψ_1 and ψ_2 are characters, then $\psi_1 < \psi_2$ if $\psi_2 = \psi_1 + \psi$ for some character ψ. This ordering is transitive and antisymmetric.

The main result below gives a description of the functional topology on $\hat{\mathfrak{A}}^{(n)}$. Let $T \in \hat{\mathfrak{A}}^{(n)}$ and $S \subset \hat{\mathfrak{A}}^{(n)}$. The following statements are equivalent ([6], Theorems 3 and 4):

(i) T lies in the functional closure of S;
(ii) $\chi_T < \psi$ for some ψ in the weak* closure of $\{\chi_s : s \in S\}$;
(iii) each φ in $\Phi(T)$ lies in the weak* closure of $\{\psi \in \Phi(T) : \|\psi\| \leq m\}$ for some integer m ($m = m(\varphi)$);
(iv) for each matrix representation S, $S \cong T$, there exist an integer k and a net $\{S^{(i)}\}$ of matrix representations, each equivalent to an element in S, with $\dim(S^{(i)}) \geq \dim(T)$ for all i and such that

$$\lim_i S_{jk}^{(i)}(a) = S_{jk}(a) \qquad (a \in \mathfrak{A}; j, k = 1, \dots, \dim T)$$

and

$$|S_{jk}^{(i)}(a)| \leq k\|a\| \qquad (j, k = 1, \dots, \dim S^{(i)}).$$

This implies ([6], Corollary 1 to Theorem 4) the following assertion: *if the Banach algebra $\mathfrak{A}$ has an identity, then each $\mathfrak{A}^{(n)}$ is compact.*

Let $\mathfrak{A} \in \pi_m$. Then there exists an n such that $\mathfrak{M} = \hat{\mathfrak{A}}^{(n)}$. For a comparison of the Gel'fand topology and the functional topology on $\mathfrak{M}$ we consider an example.

Let $\mathfrak{A}$ be the algebra of all $\mathbb{C}^{2\times 2}$-valued continuous functions f on $[0, 1]$, with $\lim_{x\to+0} f_{ij}(x) = f_{ij}(0) = 0$ for $i \neq j$. In this case $\mathfrak{M} = \hat{\mathfrak{A}}^{(2)}$, $\hat{\mathfrak{A}}_2 = (0, 1]$, and $\hat{\mathfrak{A}}_1$ is a two-point set. Let $\{T_n\}_1^\infty \subset \mathfrak{A}_2$ be the sequence defined by the equalities $T_n(f) = (f_{jn}(1/n))_{j,k=1}^2$. We first consider $\mathfrak{M}$ in the Gel'fand topology. Then $\mathfrak{M}$ is Hausdorff, but not compact: the sequence $\{T_n\}_1^\infty$ does not have limit points. Now consider $\mathfrak{M}$ in the functional topology. Then $\mathfrak{M}$ is compact, but not Hausdorff: the sequence $\{T_n\}_1^\infty$ has two limits as $n \to \infty$: $T^1(f) = f_{11}(0)$ and $T^2(f) = f_{22}(0)$.

We note that if $\hat{\mathfrak{A}} = \hat{\mathfrak{A}}_n$ for some n (for example, if $\mathfrak{A} \in \pi_n$ and it satisfies condition (α)), then the Gel'fand topology coincides with the functional topology on $\hat{\mathfrak{A}}$.

§**4.** Let $\mathfrak{A}$ be a Banach algebra. The set $\mathfrak{M}$ is naturally identified with the set of all equivalence classes of irreducible finite-dimensional representations. Let $\dim(\nu_M)$ denote the dimension of the representation space E_{ν_M}. For each

n let $\mathfrak{M}^{(n)} = \{\nu_M\colon \dim(\nu_M) \le n\}$ and $\mathfrak{M}_n = \{\nu_M\colon \dim(\nu_M) = n\}$. Suppose that $\mathfrak{A}$ satisfies condition (α). If this condition holds, then $\mathfrak{M} = \mathfrak{M}_n$.

Let $\mathfrak{M}$ have the Gel'fand topology. There is a natural homomorphism between the algebra $\mathfrak{A}$ into the algebra $C(\mathfrak{M}_n)^{n\times n}$ of all continuous $\mathbb{C}^{n\times n}$-valued functions on $\mathfrak{M}_n$:

$$\forall x \in \mathfrak{A} \qquad x \to \hat{x}(M) = \nu_M(x).$$

This is a general and obvious fact. In [7] Fell established the following result: *The algebra $\mathfrak{A}$ gives rise to an n^2-dimensional vector bundle ξ_n with base $\mathfrak{M}_n$ and fiber $\mathbb{C}^{n\times n}$.* In that article the space $\mathfrak{M}_n$ is considered in the functional topology.

THEOREM 3. *A semisimple PI-algebra satisfying condition (α) is isomorphically imbedded in the algebra $C(\mathfrak{M})^{n\times n}$.*

PROOF. Since $\mathfrak{M} = \mathfrak{M}_n$, it suffices to verify that the kernel of the natural homomorphism described above lies in the radical of $\mathfrak{A}$. Suppose that $x \in \mathfrak{A}$ and $\hat{x}(M) = 0\ \forall M \in \mathfrak{M}$, i.e., $\nu_M(x) = 0\ \forall M \in \mathfrak{M}$. Then $\forall u \in \mathfrak{A}$

$$\nu_M(e + \nu(x))\nu_M(e) + \nu_M(u)\nu_M(x) = \nu_M(e) \in G\mathbb{C}^{n\times n} \Rightarrow x \in R(\mathfrak{A}).$$

But $\mathfrak{A}$ is semisimple; consequently, $x = 0$.

We study the case when $\mathfrak{A}$ is a C^*-algebra. In [8] Fell obtained the following result: *A C^*-algebra whose irreducible representations all have dimension n is isomorphic to the algebra of all continuous sections of a certain bundle with base $\mathfrak{M}$ and fiber $\mathbb{C}^{n\times n}$.* In [9] this result was generalized to C^*-algebras whose irreducible representations all have dimension at most n. Starting from a homotopy classification of bundles, the author of [9] constructs an example of nonisomorphic C^*-algebras whose irreducible representations all have dimension n, and the representation spaces are homeomorphic. The space $\mathfrak{M}$ is considered in the functional topology in [10]. If a PI-algebra $\mathfrak{A}$ satisfies condition (α) (in this case $\mathfrak{M}$ is equipped with the Gel'fand topology), then this situation is impossible.

THEOREM 4([2]). *A C^*-PI-algebra satisfying condition (α) is isometrically $*$-isomorphic to the algebra of all continuous $n \times n$ matrix functions on the compact space $\mathfrak{M}$.*

For a proof we need

LEMMA 5. *The finite-dimensional irreducible representations are uniquely determined by their kernels (to within equivalence).*

PROOF. Suppose that ν' and ν'' are finite-dimensional irreducible representations with $\operatorname{Ker}\nu' = \operatorname{Ker}\nu''$. Consider the sets $\varphi(\nu')$ and $\varphi(\nu'')$ defined in §3. It is clear that $\varphi(\nu') = \varphi(\nu'')$. But then $\nu' \cong \nu''$ according to [6], Proposition 2. The lemma is proved.

([2])For $n = 1$ the algebra $\mathfrak{A}$ is commutative [2], and Theorem 4 is a known theorem of Gel'fand and Naĭmark ([10], §8, Theorem 1).

PROOF OF THEOREM 4. Let $\{\nu_M\}_{M\in\mathfrak{M}}$ be a sufficient set for the algebra $\mathfrak{A}$. Note that the collection $\{\nu_M\}$ contains all irreducible representations of $\mathfrak{A}$. Indeed, let ν be an irreducible representation; then $\operatorname{Ker}\nu$ is a primitive ideal. Since $\mathfrak{A}$ is a PI-algebra, $\operatorname{Ker}\nu = M \in \mathfrak{M}$; hence $\nu \cong \nu_M$ by Lemma 5. Recall that a *state* of $\mathfrak{A}$ is defined to be a positive linear functional f with $\|f\| = 1$ ([11], Definition 2.1.1). Denote by $P(\mathfrak{A})$ the set of pure states of $\mathfrak{A}$ ([11], Definition 2.5.2).

Each positive functional f determines a representation ν on a Hilbert space H and a vector $\xi \in H$ with $f(x) = (\nu(x)\xi, \xi)$ ([11], Assertion 2.4.4). Further, we assume that the standard inner product is introduced in $\mathbb{C}^{n\times n}$, turning it into a Hilbert space.

Let $f \in P(\mathfrak{A})$. Then ([11], Assertion 2.5.4) the representation ν determined by f is irreducible (according to [11], Corollary 2.8.4, we do not distinguish topologically and algebraically irreducible representations); therefore, $\nu = \nu_M$ for some $M \in \mathfrak{M}$; we write $f = f_M$. Each f_M has the form $f_M(x) = (\nu_M(x)\xi, \xi)$, where ξ runs through $\mathbb{C}^n$.

Suppose that $f_1, f_2 \in P(\mathfrak{A})$ and $f_1 \neq f_2$. We show that there exists an $x \in \mathfrak{A}$ such that $f_1(x) \neq f_2(x)$. Let $f_1(x) = (\nu_{M_1}(x)\xi_1, \xi_1)$ and $f_2(x) = (\nu_{M_2}(x)\xi_2, \xi_2)$. Since $f_1 \neq f_2$, two cases are possible:

(a) $M_1 = M_2$, but $\xi_1 \neq \lambda\xi_2$ with $|\lambda| = 1$ ([11], Proposition 2.5.7);

(b) $M_1 \neq M_2$.

Let us consider these cases.

(a) If ξ_1 and ξ_2 are linearly independent, then $\exists x \in \mathfrak{A}$: $\nu_M(x)\xi_1 = \xi_1$, $\nu_M(x)\xi_2 = 0$. Then

$$(\nu_M(x)\xi_1, \xi_1) = \|\xi_1\|^2 \neq 0 = (\nu_M(x)\xi_2, \xi_2).$$

Let $\xi_1 = \lambda\xi_2$, $|\lambda| \neq 1$; for an arbitrary $x \in \mathfrak{A}$ we have that

$$(\nu_M(x)\xi_1, \xi_1) = |\lambda|^2(\nu_M(x)\xi_2, \xi_2) \neq (\nu_M(x)\xi_2, \xi_2).$$

(b) Let $M_1 \neq M_2$. There exist operators R_1 and R_2 such that $(R_1\xi_1, \xi_1) \neq (R_2\xi_2, \xi_2)$. Then ([11], Theorem 2.8.3) $\exists x \in \mathfrak{A}$: $\nu_{M_1}(x) = R_1$, $\nu_{M_2}(x) = R_2$. We have that

$$f_1(x) = (R_1\xi_1, \xi_1) \neq (R_2\xi_2, \xi_2) = f_2(x).$$

Denote by $\mathfrak{A}_1$ the algebra of all continuous $n \times n$ matrix functions on $\mathfrak{M}$; $\mathfrak{A}$ is a C^*-algebra, and so it is semisimple. According to Theorem 3, $\mathfrak{A}$ can be imbedded isomorphically in $\mathfrak{A}_1$; denote its image by $\mathfrak{A}_2$. Clearly $\mathfrak{A}_2$ is a C^*-subalgebra of $\mathfrak{A}_1$ and $\rho(\mathfrak{A}) = \rho(\mathfrak{A}_1) = \rho(\mathfrak{A}_2)$. Using the same arguments as above, we see that $\forall f_1, f_2 \in \rho(\mathfrak{A}_1)\ \exists x \in \mathfrak{A}_2$: $f_1(x) \neq f_2(x)$, and we are in the situation of Theorem 11.1.8 in [11]. We conclude from this that $\mathfrak{A}_1 = \mathfrak{A}_2$. Accordingly, $\mathfrak{A}$ is *-isomorphic to $\mathfrak{A}_2$. To prove that this isomorphism is isometric, note that, by Theorem 2.7.3 in [11],

$$\|x\| = \sup_{M\in\mathfrak{M}} \|\nu_M(x)\| = \max_{M\in\mathfrak{M}} \|\hat{x}(M)\| = \|\hat{x}\|$$

for all $x \in A$. The theorem is proved.

BIBLIOGRAPHY

1. N. Ya. Krupnik, *Banach PI-algebras*, Investigations in Functional Analysis and Differential Equations (B. A. Shcherbakov, editor), "Shtiintsa", Kishinev, 1981, pp. 54–59. (Russian)

2. ——, *A sufficient collection of n-dimensional representations of a Banach algebra, and the n-symbol*, Funktsional. Anal. i Primenen. **14** (1980), no. 1, 63–64; English transl. in Functional Anal. Appl. **14** (1980).

3. I. N. Herstein, *Noncommutative rings*, Math. Assoc. Amer. and Wiley, 1968.

4. N. Ya. Krupnik and E. M. Shpigel', *On boundedness of a family of homomorphisms*, Mat. Issled. Vyp. 67 (1982), 69–71. (Russian)

5. Joram Lindenstrauss and Lior Tzafriri, *Classical Banach spaces*. I, Springer-Verlag, 1977.

6. J. M. G. Fell, *The dual spaces of Banach algebras*, Trans. Amer. Math. Soc. **114** (1965), 227–250.

7. ——, *Algebras and fiber bundles*, Pacific J. Math. **16** (1966), 497–503.

8. ——, *The dual spaces of C^*-algebras*, Trans. Amer. Math. Soc. **94** (1960), 365–403.

9. N. B. Vasil'ev, *C^*-algebras with finite-dimensional irreducible representations*, Uspekhi Mat. Nauk **21** (1966), no. 1(127), 135–154; English transl. in Russian Math. Surveys **21** (1966).

10. I. M. Gel'fand, D. A. Raĭkov, and G. E. Shilov, *Commutative normed rings*, Fizmatgiz, Moscow, 1960; English transl., Chelsea, New York, 1964.

11. Jacques Dixmier, *Les C^*-algèbres et leurs représentations*, 2nd ed., Gauthier-Villars, Paris, 1969.

Translated by H. H. McFADEN

Amer. Math. Soc. Transl.
(2) Vol. **142**, 1989

On the Solvability of Nonlinear Closed Operator Equations

UDC 513.88

D. S. DZHUMABAEV

Abstract. An approach to the investigation of nonlinear operator equations is presented. Existence theorems are obtained for solutions of these equations, of which one generalizes a theorem of Hadamard.

We consider the nonlinear operator equation

$$A(x) = 0, \tag{1}$$

where $x \in B_1$, $A(x) \in B_2$, and each B_i is a Banach space with norm $\|\cdot\|_i$, $i = 1, 2$. Denote by $D(A)$ and $R(A)$ the domain and range of A, respectively. We take a point $x^0 \in D(A)$ and form the sets

$$S(x^0, r) = \{x \in B_1 \colon \|x - x^0\|_1 \le r\},$$
$$U^0 = \{x \in D(A) \colon \|A(x)\|_2 \le \|A(x^0)\|_2 = u^0\}, \qquad \Omega = S(x^0, r) \cap U^0.$$

Assume that the operator A is closed on Ω ([1], Chapter XII, §1.6). As is known [1], [2], iteration methods allowing one to find the solution itself under sufficient conditions for its existence make use of certain linearizations of the nonlinear operator. Therefore, we introduce the concept of a linearizer of an operator A at a point $\hat{x} \in D(A)$.

Definition. A linear operator $C \colon B_1 \to B_2$ is called a *linearizer* of an operator A at a point $\hat{x} \in D(A)$, if $D(A) \subseteq D(C)$ and there exist numbers $\epsilon \ge 0$ and $\delta > 0$ such that if $\|x - \hat{x}\|_1 < \delta$, $x \in D(A)$, then

$$\|A(x) - A(\hat{x}) - C(x - \hat{x})\|_2 \le \epsilon \|x - \hat{x}\|_1.$$

Remark 1. If $C \in L(B_1, B_2)$ is the Fréchet derivative of A at a point $\hat{x} \in D(A)$, then it is also a linearizer. However, in contrast to the case of

1980 *Mathematics Subject Classification* (1985 *Revision*). Primary 47H17, 47H15.
Translation of Izv. Akad. Nauk Kazakh. SSR Ser. Fiz.-Mat. **1984**, no. 1, 31–34; MR **85g**:47084.

the Fréchet derivative, the definition of a linearizer does not require: a) the boundedness of the operator C, or b) the dependence of the number ϵ on δ ($\epsilon(\delta) \to 0$ as $\delta \to 0$ for the Fréchet derivative).

REMARK 2. While the Fréchet derivative of an operator A is uniquely determined, there can be infinitely many linearizers of this operator.

Distinctive advantages of linearizers make it possible to extend the domain of application of iteration methods for solving nonlinear operator equations.

It will be assumed in what follows that the domain $D(C)$ and the range $R(C)$ of a linearizer C_x do not depend on $x \in D(A)$, and, furthermore, $D(A) = D(C) \cap S(x^0, r)$ and $R(A) \subseteq R(C)$. We have the following main result.

THEOREM 1. *Suppose that at each point $x \in \Omega$ the operator A has a linearizer C_x with constants ϵ_x and δ_x such that:* 1) *C_x is a one-to-one mapping of $D(C)$ onto $R(C)$, and $\|C_x^{-1}\| \leq \gamma_x \leq \overline{\gamma}$;* 2) *$\epsilon_x \cdot \gamma_x \leq \Theta < 1$; and* 3) *$(\gamma_x/\delta_x) \cdot \|A(x)\|_2 \leq K$. If*

$$\frac{\overline{\gamma}}{1-\Theta} \cdot \|A(x^0)\|_2 < r$$

then (1) *has a solution $x^* \in \Omega$, and the iteration process*

$$x^{(n+1)} = x^{(n)} - \frac{1}{\alpha} C_{x^{(n)}}^{-1}\{A(x^{(n)})\} \tag{2}$$

converges to it, where $\alpha = \max\{1, K\}$, $n = 0, 1, 2, \ldots$.

PROOF. Take the initial point x^0, and let the subsequent approximations be determined by the iteration process (2). Let $C_n = C_{x^{(n)}}$, $\epsilon_n = \epsilon_{x^{(n)}}$, $\gamma_n = \gamma_{x^{(n)}}$, and $\delta_n = \delta_{x^{(n)}}$. Since $x^0 \in D(C)$, $C_0^{-1}\{A(x^0)\} \in D(C)$, and $D(C)$ is a linear manifold, it follows that the $x^{(1)}$ determined from (2) with $n = 0$ belongs to $D(C)$. On the other hand, considering the estimate

$$\|x^{(1)} - x^0\|_1 \leq \frac{1}{\alpha}\gamma_0 \cdot u^0 < \frac{\overline{\gamma}}{1-\Theta}u^0 \leq r,$$

we have that $x^{(1)} \in D(A)$ and

$$A(x^{(1)}) = A(x^{(1)}) - A(x^0) - C_0(x^{(1)} - x^0) + \left(1 - \frac{1}{\alpha}\right) A(x^0).$$

By condition 3), $\delta_0 \geq \gamma_0 u^0/K \geq \gamma_0 u^0/\alpha$; consequently, $\|x^{(1)} - x^0\|_1 \leq \delta_0$, and, by the definition of the linearizer,

$$\|A(x^{(1)})\|_2 \leq \epsilon_0 \cdot \|x^{(1)} - x^0\|_1 + \left(1 - \frac{1}{\alpha}\right) \cdot \|A(x^0)\|_2 \leq \frac{\epsilon_0\gamma_0}{\alpha}\|A(x^0)\|_2$$
$$+ \left(1 - \frac{1}{\alpha}\right) \cdot \|A(x^0)\|_2 = \left[1 - \frac{1}{\alpha}(1 - \epsilon_0\gamma_0)\right] \cdot \|A(x^0)\|_2.$$

Here $\beta_0 = [1 - (1 - \epsilon_0\gamma_0)/\alpha] < 1$; therefore $x^{(1)} \in \Omega$. Similarly,

$$\|x^{(n)} - x^{(n-1)}\|_1 \leq \frac{\gamma_{n-1}}{\alpha} \cdot \|A(x^{(n-1)})\|_2, \qquad \|A(x^{(n)})\|_2 \leq \beta_{n-1} \cdot \|A(x^{(n-1)})\|_2,$$

where

$$\beta_{n-1} = 1 - \frac{1}{\alpha}(1 - \epsilon_{n-1} \cdot \gamma_{n-1}),$$

and $x^{(n)}$ belongs to the set Ω. According to requirement 2) of the theorem, $\beta_{n-1} \leq \overline{\beta} = 1 - (1 - \theta)/\alpha$. This and the denseness of B_1 give us that $\{x^{(n)}\}$ $(n = 0, 1, \dots)$ is a Cauchy sequence, and it converges to $x^* \in B_1$. Since A is closed, we now get that $x^* \in \Omega$ and $A(x^*) = 0$ from the relations $\|x^{(n)} - x^*\|_1 \to 0$ and $\|A(x^{(n)})\|_2 \to 0$ as $n \to \infty$. The theorem is proved.

As noted above, the constants ϵ_x and δ_x do not depend on each other in general. We consider the case when for a given $\delta > 0$ there is an $\epsilon(\delta)$ independent of x.

THEOREM 2. *Suppose that at each point $x \in \Omega$ and for each $\delta \in (0, h)$ the operator A has a linearizer C_x with constants $\delta > 0$ and $\varepsilon(\delta) \geq 0$ satisfying the following conditions:*

1) *C_x^{-1} exists on $R(C)$, and $\|C_x^{-1}\| \leq \gamma$,*

2) $\lim_{\delta \to 0} \epsilon(\delta) = 0$.

Then (1) *has a solution $x^* \in \Omega$ if the following inequality holds:*

3) $\gamma \cdot \|A(x^0)\|_2 < r$.

PROOF. We again use the iteration process (2) for some $\alpha \geq 1$ whose value we determine below. Take any $\delta > 0$ and an $\epsilon(\delta)$ corresponding to it. As proved in the preceding theorem, if we choose an $\alpha \geq \gamma u^0/\delta$, then the successive approximations $x^{(n)}$ converge to $x^* \in U^0$ as $n \to \infty$, and

$$\|x^* - x^0\|_1 \leq \frac{1}{1 - [1 - \frac{1}{\alpha}(1 - \epsilon\gamma)]} \cdot \frac{\gamma}{\alpha} \cdot \|A(x^0)\|_2 = \frac{\gamma}{1 - \epsilon\gamma} \cdot \|A(x^0)\|_2.$$

Inequality 3) implies that there is an $\tilde{\epsilon} > 0$ such that $\gamma\|A(x^0)\|_2 < r(1 - \tilde{\epsilon} \cdot \gamma)$. On the basis of 2) we now find a $\tilde{\delta} > 0$ such that $\epsilon(\tilde{\delta}) < \tilde{\epsilon}$. It remains to take $\tilde{\alpha} \geq \gamma u^0/\tilde{\delta}$ and to use the iteration process (2) with $\alpha = \max\{1, \tilde{\alpha}\}$. In this case the successive approximations $x^{(n)}$ converge to $x^* \in \Omega$ as $n \to \infty$. The theorem is proved.

If A has a uniformly continuous Fréchet derivative on Ω, then we use it for C_x, and we take the constants in the definition of the Fréchet derivative for δ and $\epsilon(\delta)$. Then, by Theorem 2, the bounded invertibility of C_x and inequality 3) imply that there is a solution $x^* \in \Omega$. Therefore, Theorem 2 is a generalization of the local theorem of Hadamard ([2], Theorem 5.3.11) to unbounded operator equations.

As an example we consider a nonlinear boundary value problem for an ordinary differential equation:

$$\frac{dz}{dt} = f(t, z), \qquad t \in [-T, T], \quad z(-T) = z(T). \tag{3}$$

Assume that in some interval $[z^0 - r,\ z^0 + r]$ and for each fixed $t \in [-T, T]$ the function $f(t, z)$ has a continuous derivative with respect to z, and $|\partial f/\partial z| \geq q(t) > 0$. We take the space B_1 to be the set of continuous functions $z(t)$ on $[-T, T]$ such that $z(-T) = z(T)$, with the norm

$$\|z(t)\| = \max_{t\in[-T,T]} |z(t)|.$$

Then problem (3) is equivalent to the equation

$$A(z) = 0, \qquad \text{where} \qquad z \in B_1, \qquad A(z) \equiv \frac{1}{q(t)}\left[\frac{dz}{dt} - f(t, z)\right].$$

Let $S(z^0, r) = \{z \in B_1 : \|z - z^0\| \leq r\}$, and let H_1 be the set of functions $z \in B_1$ having a continuous derivative on $[-T, T]$. The operator A is closed in its domain $D(A) = H_1 \cap S(z^0, r)$, and the linear operator

$$C_{\hat{z}} \equiv \frac{1}{q(t)}\left[\frac{d}{dt} - f_z'(t, \hat{z}(t))\right]$$

with domain $D(C) = H_1 \supset D(A)$ is a linearizer of A at the point $\hat{z}(t) \in D(A)$.

It can be shown by the method of parametrization [3] that the linear boundary value problem

$$\frac{dz}{dt} = f_z'(t, \hat{z}(t)) \cdot z + q(t) \cdot F(t), \qquad z(-T) = z(T),$$

has a unique solution $\overline{z}(t)$, and it satisfies the inequality

$$\|\overline{z}(t)\| \leq \left\|\frac{q(t) \cdot F(t)}{f_z'(t, \hat{z}(t))}\right\| \leq \|F(t)\|,$$

with $F(t)$ continuous on $[-T, T]$, which ensures that $C_{\hat{z}}$ is invertible and $\|C_{\hat{z}}^{-1}\| \leq 1$. Then (3) has a solution $z^*(t) \in S(z^0, r) \cap H_1$ if $\|A(z^0)\| = \|\frac{f(t,z^0)}{q(t)}\| < r$, by Theorem 2.

BIBLIOGRAPHY

1. L. V. Kantorovich and G. P. Akilov, *Functional analysis*, 2nd rev. ed., "Nauka", Moscow, 1977; English transl., Pergamon Press, 1982.

2. J. M. Ortega and W. C. Rheinboldt, *Iterative solution of nonlinear equations in several variables*, Academic Press, 1970.

3. D. S. Dzhumabaev, *On a method for investigating ordinary differential equations*, Izv. Akad. Nauk Kazakh. SSR Ser. Fiz.-Mat. **1982**, no. 3, 1–4. (Russian)

Translated by H. H. McFADEN

Amer. Math. Soc. Transl.
(2) Vol. **142**, 1989

On the Convergence of a Modification of the Newton–Kantorovich Method for Closed Operator Equations

UDC 513.88

D. S. DZHUMABAEV

Abstract. Conditions are obtained for convergence of certain modifications of the Newton–Kantorovich method for closed (unbounded) nonlinear operator equations.

In [1] we introduced the concept of a linearizer, which generalizes the Fréchet and Gâteaux derivatives, and we proposed a certain method for proving convergence of iteration processes for closed operator equations.

In the present article this method is applied to the investigation of convergence of certain modifications of the Newton–Kantorovich method ([2], [3]) for an equation of the form

$$A(x) \equiv Cx + F(x) = 0, \tag{1}$$

where $C : X \to Y$ is a closed linear operator, $F : X \to Y$ is a continuous operator, and X and Y are Banach spaces with respective norms $\|\cdot\|_1$ and $\|\cdot\|_2$. Denote by $D(A)$ and $R(A)$ the domain and range of an operator A. Assume that F has a Fréchet derivative in some domain containing the ball $\overline{S}(x^0, r) = \{x \in X : \|x - x^0\|_1 \le r\}$, $x^0 \in D(C)$, and $R(C + F'(x)) = Y$ for $x \in S(x^0, r)$. Then in $D(A) = D(C) \cap \overline{S}(x^0, r)$ the operator A has the linearizer $C_1(x) = C + F'(x)$, and $D(C_1) = D(C) \cap X = D(C)$.

THEOREM 1. *Assume the following conditions hold.*

1) *For all $x \in D(A)$ the linearizer $C_1(x)$ is boundedly invertible, and $\|C_1^{-1}(x)\|_{L(Y,X)} \le \gamma$.*

2) $\|F'(x) - F'(y)\|_{L(X,Y)} \le L \cdot \|x - y\|_1, \quad x, y \in \overline{S}(x^0, r).$

1980 *Mathematics Subject Classification* (1985 *Revision*). Primary 47H17, 47H15, 49D15.

Translation of Izv. Akad. Nauk. Kazakh. SSR Ser. Fiz.-Mat. **1984**, no. 3, 25–28; MR **85m**: 47068.

0065-9290/89 \$1.00 + \$.25 per page

3) $\frac{m}{L\gamma} + \gamma \frac{b_m}{b_0} \|A(x^0)\|_2 \sum_{s=0}^{\infty} (b_m)^{2^s - 1} < r$, *where* $b_0 = \frac{L}{2}\gamma^2 u_0$, $u_0 = \|A(x^0)\|_2$, $\beta_k = 1 - \frac{1}{4b_{k-1}}$, $b_k = \beta_k \cdot b_{k-1}$, $k = 1, 2, \dots, m$,
where m *is a nonnegative number such that* $b_m < 1$ *and* $b_{m-1} \geq 1$.

Then the damped Newton–Kantorovich method

$$x^{(k+1)} = x^{(k)} - \frac{1}{\alpha_k}[C + F'(x^{(k)})]^{-1}[Cx^{(k)} + F(x^{(k)})], \quad k = 0, 1, 2, \dots, \tag{2}$$

where $\alpha_k = 2b_k$ *for* $k = 0, \dots, m-1$ *and* $\alpha_k = 1$ *for* $k = m, m+1, \dots$, *converges to a solution of* (1).

PROOF. Take the center x^0 of the ball for the zero approximation. Since $x^0 \in D(C)$, $[C + F'(x^0)]^{-1}[Cx^0 + F(x^0)] \in D(C)$, and $D(C)$ is a linear manifold, the $x^{(1)}$ determined by (2) with $k = 0$ lies in $D(C)$. By an assumption of the theorem,

$$\|x^{(1)} - x^0\|_1 \leq \frac{1}{\alpha_0} \cdot \gamma \cdot \|A(x^0)\|_2 = \frac{1}{2b_0} \cdot \gamma \cdot u_0 = \frac{2\gamma u_0}{2L\gamma^2 u_0} = \frac{1}{L\gamma} < r.$$

Therefore, $x^{(1)} \in D(A)$. Using the estimate

$$\|F(x^{(1)}) - F(x^0) - F'(x^0)(x^{(1)} - x^0)\|_2 \leq \frac{L}{2}\|x^{(1)} - x^0\|_1^2,$$

we get from the equality

$$A(x^{(1)}) = A(x^{(1)}) - A(x^0) - [C + F'(x^0)](x^{(1)} - x^0) + \left(1 - \frac{1}{\alpha_0}\right) A(x^0)$$

that

$$\begin{aligned}
\|A(x^{(1)})\|_2 &\leq \frac{L}{2}\|x^{(1)} - x^0\|_1^2 + \left(1 - \frac{1}{\alpha_0}\right)\|A(x^0)\|_2 \\
&\leq \frac{L}{2}\frac{\gamma^2}{\alpha_0^2} u_0 \cdot \|A(x^0)\|_2 + \left(1 - \frac{1}{\alpha_0}\right)\|A(x^0)\|_2 \\
&= \left[1 - \frac{1}{\alpha_0}\left(1 - \frac{L}{2}\frac{\gamma^2 u_0}{\alpha_0}\right)\right] \cdot \|A(x^0)\|_2 = \beta_1 \cdot \|A(x^0)\|_2.
\end{aligned}$$

It is established similarly that $x^{(k)}$ belongs to $D(C)$ and

$$\begin{aligned}
\|x^{(k)} - x^{(k-1)}\|_1 \leq \frac{1}{2b_{k-1}} \cdot \gamma \cdot \|A(x^{(k-1)})\|_2 &\leq \frac{\gamma}{2b_{k-1}} \cdot \frac{b_{k-1}}{b_0} \cdot u_0 \\
&= \frac{2\gamma u_0}{2\gamma^2 L u_0} = \frac{1}{L\gamma}, \\
\|A(x^{(k)})\|_2 \leq \beta_k \cdot \|A(x^{(k-1)})\|_2, \quad & k = 2, 3, \dots, m.
\end{aligned}$$

The subsequent approximations are now determined by the basic Newton–Kantorovich method:

$$x^{(m+s+1)} = x^{(m+s)} - [C + F'(x^{(m+s)})]^{-1}[Cx^{(m+s)} + F(x^{(m+s)})], \quad s = 0, 1, 2, \dots. \tag{3}$$

Then $\|x^{(m+1)} - x^{(m)}\|_1 \leq \gamma\|A(x^{(m)})\|_2$, and, by (3),

$$\|x^{(m+1)} - x^0\|_1 \leq \|x^{(m+1)} - x^{(m)}\|_1 + \|x^{(m)} - x^0\|_1 \leq \frac{\gamma \cdot b_m}{b_0} \cdot \|A(x^0)\|_2 + \frac{m}{L\gamma} < r.$$

The fact that $x^{(m+1)}$ and all the subsequent approximations $x^{(m+s)}$ belong to $D(C)$ follows from the arguments given at the beginning of the proof.

In what follows we only check the inequality $\|x^{(m+s)} - x^0\|_1 < r$. It follows from the equality

$$A(x^{(m+1)}) = A(x^{(m+1)}) - A(x^{(m)}) - [C + F'(x^{(m)})](x^{(m+1)} - x^{(m)})$$

that

$$\|A(x^{(m+1)})\|_2 \leq \frac{L}{2}\|x^{(m+1)} - x^{(m)}\|_1^2 \leq \frac{L}{2}\gamma^2\|A(x^{(m)})\|_2^2 \leq \frac{L}{2}\gamma^2\frac{b_m}{b_0}u_0 \cdot \|A(x^{(m)})\|_2 = b_m \cdot \|A(x^{(m)})\|_2.$$

If we assume that

$$\|A(x^{(m+s)})\|_2 \leq b_m^{2^s-1} \cdot \|A(x^{(m)})\|,$$

then

$$\|A(x^{(m+s+1)})\|_2 \leq \frac{L}{2}\gamma^2\|A(x^{(m+s)})\|_2^2 \leq \frac{L}{2}\gamma^2 b_m^{2^{s+1}-2} \cdot \|A(x^{(m)})\|_2^2 \leq b_m^{2^{s+1}-2} \cdot \frac{L}{2}\gamma^2\frac{b_m}{b_0}u_0\|A(x^{(m)})\|_2 = b_m^{2^{s+1}-1} \cdot \|A(x^{(m)})\|_2.$$

Therefore, we have the basic relations

$$\|A(x^{(m+s)})\|_2 \leq b_m^{2^s-1} \cdot \frac{b_m}{b_0}u_0, \qquad \|x^{(m+s+1)} - x^{(m+s)}\|_1 \leq \gamma b_m^{2^s-1} \cdot \frac{b_m}{b_0}u_0, \qquad s = 0, 1, 2, \ldots,$$

which imply that $\|A(x^{(m+s)})\|_2$ converges to zero and $x^{(m+s)}$ converges to an $x^* \in \overline{S}(x^0, r)$ as $s \to \infty$, and $x^{(m+s)} \in D(A)$, $s = 1, 2, \ldots$. The closedness of C and the continuity of F in $D(A)$ ensure the closedness of A in $D(A)$, and thus $x^* \in D(A)$ and $A(x^*) = 0$, i.e., $Cx^* + F(x^*) = 0$.

The theorem is proved.

REMARK. If in Theorem 1 we assume that C is the zero operator and the number m is equal to zero, then we get a known theorem of Mysovskiĭ on convergence of the basic Newton–Kantorovich method.

Conditions for convergence of the modified Newton–Kantorovich method $x^{(n+1)} = x^{(n)} - [C + F'(x^0)]^{-1}[Cx^{(n)} + F(x^{(n)})]$, $n = 0, 1, 2, \ldots$, are obtained in [4] by applying the contraction mapping principle to the equivalent equation

$$x = [C + F'(x^0)]^{-1}[F'(x^0)x - F(x)].$$

Difficulties connected with checking the conditions (3) arise in the concrete use of theorems analogous to Theorem 1. Since the numbers L, γ, and m are not known in advance and depend on r, it is necessary to solve fairly complicated one-dimensional inequalities. The following theorem simplifies (3), reducing it to the verification of a relation between γ and r for the chosen x^0.

THEOREM 2. *Assume the following conditions hold.*

1) *For all* $x \in D(A)$ *the linearizer* $C + F'(x)$ *is boundedly invertible, and* $\|[C+F'(x)]^{-1}\|_{L(Y,X)} \leq \gamma$.

2) *The Fréchet derivative* $F'(x)$ *is uniformly continuous in* $\overline{S}(x^0, r)$.

3) $\gamma \cdot \|A(x^0)\|_2 < r$.

Then there exist numbers $\alpha_n \geq 1$, $n = 0, 1, \dots$, *such that the iteration process* (3) *converges to an isolated solution* $x^* \in D(A)$ *of* (1). *Further, beginning with some* k^0, *we can take* α_n $(n \geq k^0)$ *equal to* 1, *and the convergence is superlinear.*

PROOF. It follows from (3) that there exist numbers $\varepsilon_0 > 0$ and $\varepsilon > 0$ such that $\varepsilon_0\gamma < 1$ and $\frac{\gamma\|A(x^0)\|_2}{(1-\varepsilon_0\gamma)} + \varepsilon < r$. By (2), there exists a $\delta_0 > 0$ such that $\|F'(x) - F'(y)\|_{L(X,Y)} < \varepsilon_0$ if $\|x - y\|_1 \leq \delta_0$. Using the mean value theorem, we have that

$$\begin{aligned}\|A(x) - A(y) - [C + F'(x)](x - y)\|_2 &= \|F(x) - F(y) - F'(x)(x - y)\|_2 \\ &= \left\| \int_0^1 \{F'[x + t(y - x)] - F'(x)\}\, dt(x - y) \right\|_2 < \varepsilon_0 \cdot \|x - y\|_1\end{aligned}$$

if $\|x - y\|_1 < \delta_0$, $x, y \in \overline{S}(x^0, r)$. The number α_0 is chosen equal to $\max\{1, \gamma u_0/\delta_0\}$, where $u_0 = \|A(x^0)\|_2$. For this value of α_0 we have that $\|x^{(1)} - x^0\|_1 \leq \delta_0$ and

$$\|A(x^{(1)})\|_2 \leq \left[1 - \frac{1}{\alpha_0}(1 - \varepsilon_0\gamma)\right] \cdot \|A(x^0)\|_2 = \beta_1 \cdot u_0.$$

Since $\beta_1 < 1$, there exists a positive integer k^0 satisfying the inequality $(\gamma \cdot u_0/(1 - \varepsilon_0\gamma)) \cdot \beta_1^k \leq \varepsilon$. Then take α_n equal to α_0 when $n = 1, \dots, k^0 - 1$, and equal to 1 when $n = k^0$, $k^0 + 1$, $k^0 + 2, \dots$.

It is now not hard to establish the estimates

$$\|x^{(n)} - x^{(n-1)}\|_1 \leq \frac{1}{\alpha_0}\gamma \cdot \|A(x^{(n-1)})\|_2, \qquad \|A(x^{(n)})\|_2 \leq \beta_1 \cdot \|A(x^{(n-1)})\|_2,$$
$$n = 1, 2, \dots, k^0,$$
$$\|x^{(k)} - x^{(k-1)}\|_1 \leq \gamma \cdot \|A(x^{(k-1)})\|_2, \qquad \|A(x^{(k)})\|_2 \leq \varepsilon_0 \cdot \gamma\|A(x^{(k-1)})\|_2,$$
$$k = k^0 + 1, k^0 + 2, \dots;$$

it follows from these and the principle for choosing the numbers ε_0, ε, and k^0 that all the sequences $x^{(n)}$ lie in $D(A)$, $x^{(n)} \to x^*$, $\|A(x^{(n)})\|_2 \to 0$ as $n \to \infty$, and x^* is an interior point of $\overline{S}(x^0, r)$. Considering that the operator A is closed, we get that $x^* \in D(A)$ and $A(x^*) \equiv Cx^* + F(x^*) = 0$. We take the number $\bar{\delta}$ so that $S(x^*, \bar{\delta}) \subset \overline{S}(x^0, r)$.

Suppose that there is another solution y^* in $D(A) \cap S(x^*, \delta_1)$, where $\delta_1 = \min\{\delta_0, \bar{\delta}\}$. Then the estimate

$$\begin{aligned}\|x^* - y^*\|_1 &= \|[C + F'(x^*)]^{-1}\{[C + F'(x^*)](x^* - y^*) - A(x^*) + A(y^*)\}\|_1 \\ &< \gamma\varepsilon_0\|x^* - y^*\|_1\end{aligned}$$

implies that $x^* = y^*$. Therefore, x^* is an isolated solution of (1). For $n \geq k^0$ we have the inequality

$$\|x^{(n+1)} - x^{(n)}\|_1 \leq \gamma \varepsilon_n \cdot \|x^{(n)} - x^{(n-1)}\|_1,$$

where $\varepsilon_n \to 0$ as $n \to \infty$. This gives us that the convergence of the iteration process (2) is superlinear.

The theorem is proved.

BIBLIOGRAPHY

1. D. S. Dzhumabaev, *On the solvability of nonlinear closed operator equations*, Izv. Akad. Nauk Kazakh. SSR Ser. Fiz.-Mat. **1984**, no. 1, 31–34; English transl. in this volume.

2. L. V. Kantorovich and G. P. Akilov, *Functional analysis*, 2nd rev. ed., "Nauka", Moscow, 1977; English transl., Pergamon Press, 1982.

3. J. M. Ortega and W. C. Rheinboldt, *Iterative solution of nonlinear equations in several variables*, Academic Press, 1970.

4. M. N. Yakovlev, *Estimates of solutions of two-point problems for systems of first-order integro-differential equations, and the method of lines*, Zap. Nauchn. Sem. Leningrad. Otdel. Mat. Inst. Steklov. (LOMI) **102** (1980), 156–173; English transl. in J. Soviet Math. **22** (1983), no. 2.

Translated by H. H. McFADEN

Amer. Math. Soc. Transl.
(2) Vol. **142**, 1989

Σ-Programming

UDC 519.685

S. S. GONCHAROV AND D. I. SVIRIDENKO

The goal of this article is to describe a new programming concept which relies on contemporary ideas and methods of mathematical logic. This concept, which bears the name "Σ-programming", is based on the fact that computable functions admit formal descriptions. In Σ-programming, formulas of a special type (which are called Σ-formulas) in a language for computation consisting of first-order predicates are considered as constituting descriptions of this kind.

§1. On some trends in program development

Within the set of problems which are treated in programming, those of the following type enjoy the greatest popularity:

—the quality, in particular, the reliability and correctness of program specifications;

—productivity of the programmer's work;

—"intellectualization" of programming systems and computers.

It is not hard to see that all these and other problems of contemporary programming are closely connected to one another, and thus attempts to resolve them separately, in isolation from each other, appear unfounded. In other words, a complex, conceptually unified approach is to be preferred. The search for such an approach, and its implementation, naturally depends on how the investigator conceives of the content of notions such as information, or an information processing activity. Furthermore, such questions as "What is such a problem?" or "What does it mean to solve a problem?", etc. will be no less important. All this and the related complex of problems and questions arise in practice in any area of programming. We will consider some of these.

1980 *Mathematics Subject Classification* (1985 *Revision*). Primary 68N05; Secondary 68N15.
Translation of Vychisl. Sistemy Vyp. 107 (1985), 3–29; MR **87g**:68009.

1. *The construction of programs.* The current view of the process of program development is connected above all with a division of this process into definite stages ("life cycles" of the program): the specification stage, the design stage, the stage in which the initial text of the program is coded, etc. A similar level of structuralization is determined mainly by the complexity and the scale of the programs under construction, which in general leads to the problem of the adequacy of the programs for the problems which they were intended to solve. The adequacy of programs (in other words their "correctness" or "appropriateness") is necessary since we need *confidence* in the correctness and reliability of their operation; we wish to *verify* programs—otherwise, why would we need them? Therefore, methods for proving such adequacy are needed. The choice among various methods naturally depends on our concept of a *specification*, as well as our view of the *design process* of programs and its relation to the specification, what we mean by a *program*, and how its *implementation* proceeds. If such methods are compatible with a natural formalism, then on the one hand there is the possibility of automation, either of the whole process of the construction of programs according to their specifications (automatic program synthesis) or of separate fragments (automatic design); and on the other hand, such a formalism, for its effective realization, may require completely new organizational principles for computation processes and computers. The latter might in general lead to a modification of our representation of what a specification, design, or encoding of a program is, and even what a program itself is.

2. *Data structures.* When we speak of the automation of a certain activity of information processing, this implies that it is a matter above all of our ability to solve problems such as an adequate representation of the information in the form of a *data structure* and an adequate representation of the process to be automated in the form of an *algorithm* operating on the data structure. We remark that in speaking of data structures and algorithms, we emphasize involuntarily, as it were, the distinction between these two concepts. It is necessary to mention that the point of view which takes the given distinction as absolute, and according to which: "a program = an algorithm + a data structure" is extremely popular in programming—and there is a reason for this. For example, thanks to the separation of the data structures from the algorithms' operation on them, it is possible to begin the design of a program, let us say, without an exact representation of the requirements in the data structures, that is, of how the data structures will be completely and exactly defined. Furthermore, a change in the data structures is an extremely common method of improving a program. Here, only modifications in the last stages of program design are concerned. At the same time, there is an approach in programming to the creation of programs which puts the data aside, i.e., the design is not guided by the algorithm, but by the data structure (for example, programming technology). But even here a clear separation of the concepts "data structure" and "algorithm" occurs. Only

recently, especially in connection with the emergence of the concept of "abstract data types", was it gradually realized that similarly the separation of the concepts of "data structure" and "algorithm" has an artificial rather than an intrinsic character. These ideas quickly react on the roles which are played by programming constructs (and even a single construct) in the computation process—everything depends on the way that these constructs are employed. "Abstract data types": it is in fact this context which controls the utilization of objects and in which the utilization of one or another of them proceeds.

It is necessary to remark that the concept of "abstract data types" allows us to look at many aspects of programming with a more or less unified point of view. As a consequence the level of linguistic constructions and their diversity is raised, and programming languages may acquire a more declarative character. In other words, as the level of automation of information processes becomes higher, the process of problem orientation of the linguistic methods becomes simpler (from the user's point of view). But the concept of "abstract data types" requires the creation of adequate system-technical methods of data processing, i.e., new computing machines based on models of computation different from the traditional von Neumann model.

In the area of "abstract data types" there is currently much activity to be seen dealing with measurements of scientific experiments. In the framework of this concept, at present many different approaches have been developed (initial, terminal, model-theoretic, categorical, axiomatic, etc.). It is necessary to mention that they are not all worked out methodologically and mathematically. The question of the systematic and possibly definitive unification of the knowledge accumulated in this area remains on the agenda.

3. *The description of data bases and operations on them.* The solution of many problems often requires a huge mass of information. It is therefore essential to use computers in the solution of such problems. But here the problem of the machine representation of huge quantities of information arises, as well as its organization, structuralization, and the provision of simple and effective access to it, and the possibility of continual storage and updating of the information. Similar problems arise in the construction of control systems for complex technical, military, economic, and social systems, and in the creation of information-retrieval, query-response, and other "intelligent" systems, in particular of expert systems. For the last type of system, problems connected to their "intellectualization" are characteristic: problems of the organization of a dialog, interaction in natural languages, and problems of "intellectualization" of data processing (logical deduction, mechanical theorem proving, hypothesis generation, planning the order of computation, pattern recognition, etc.). Clearly it is necessary to solve all these problems as a complex. Specific progress in this area is connected with the so-called *relational* approach. More hope is placed in abstract data. The most promising approach (and what is principally a synthetic one) takes the form of *logical programming.* But also here the problem of the effective re-

alization of all these approaches arises, i.e., of supporting everything with computational methods. A similar problem acquires an even more complex conceptual and technical character in work with the so-called *distributed* data bases.

4. *Computer architecture.* We have already emphasized repeatedly that the realization of new concepts of the representation and processing of information requires computational methods which are adequate to these concepts. In other words, an approach to programming can count on positive results only if it involves not only simple programming, but also the technical aspects, i.e., aspects which are connected with the organization of computation processes and the architecture of computing systems.

Most contemporary ideas in programming are some sort of realization of the von Neumann model of computation. But increasingly it appears that the constriction in this model (in Backus' apt phrase, "the von Neumann bottleneck") places essential obstacles in the path of an effective realization of these ideas. New machines are needed; and they are appearing. Thus, for example, the latest realization of the idea of *functional* (or, in another terminology, *applicative*) programming led to the appearance of designs in which not only functional languages were developed, but also "functional" machines—"stream" machines, a different form of LISP-machine, "reduction" machines, and the like. As an example of such a unified conceptual design we may cite the project under Turner, in which context the language SASL (Saint Andrews Static Language) and the machine KRC (Kent Recursive Calculator) were developed. Among the functional machines we may recall such "reduction" machines as SKIM, ZAPP, ALICE (England), SKY (Sweden), GMD (F.R.G.), and "functional" machines realizing the ideas of Backus: ARVIND, the FFP-machine (U.S.A.), etc. In the Japanese project for the creation of fifth-generation computers the development of "functional" machines is also envisaged. In the same project the construction of computers which incorporate the concepts of logic programming (PROLOG-machines) is envisaged. Not long ago, the creation of an experimental PROLOG-machine, PAROLOG, was announced. The development of "relational" computers and computers which incorporate the concept of "abstract data types" is actively pursued. In the development of "functional", "logical", and other computational machines various forms of parallel information processing are actively utilized (for example OR- and AND-parallelism). And although similar developments were possible in great part thanks to advances in current technology, we must mention that it appears to us as observers that completely new trends in programming—the appearance of new programming concepts—are determinative in the creation of the technical basis for programming.

To conclude our brief analysis of the state of affairs in some areas of programming, it is impossible to omit one important circumstance: there is a clear conceptual similarity between the approaches recalled above. As a corroboration of this we see, for example, the possibility of successfully realizing the ideas of logic programming on "reduction" machines. At the same time, functional programming may be regarded as a special (equational) case of logic programming. It is also possible to discuss the relational approach. The most clearly visible connection lies in the concepts of "abstract data types" and the relational approach. In the context of the former concept it could be extremely interesting to pose the problem of the synthesis of programs according to specifications. The reason for such a conceptual affinity is clear: all these approaches are of a logico-mathematical nature. Thus we discern an active penetration of the ideas and methods of mathematical logic into programming. Furthermore, we may predict that mathematical logic will become the synthesizing point of departure which the current methodologies, theories, and technologies of programming do not provide. This remark leads us to hope that it will also be possible to have mathematical-logical concepts of mechanical information processing which will enable us to regard all the problems discussed earlier from a unified standpoint. A scientific search is needed, in its own way an experiment in the given direction. The present article treats just such an investigation.

§2. The methodology of Σ-programming

The basic concept of Σ-programming is the concept of a "problem". A problem in Σ-programming is represented by a *specification.* Among the specifications for problems a special class of so-called *Σ-specifications* is distinguished, which may be "fulfilled". It is assumed that the "fulfillment" of a Σ-specification is carried out by some device: a *Σ-machine.*([1]) Among the "fulfillable" specifications a special place is taken by the Σ-specifications, whose "fulfillment" by Σ-machines is carried out in a sufficiently effective way. We will call such Σ-specifications *Σ-programs.* Σ-specifications and Σ-programs do not differ syntactically. The choice of the Σ-programs is carried out on purely pragmatic considerations, with the only criterion being "effective fulfillment". Since in this article the latter notion is never made precise, we will not distinguish Σ-specifications and Σ-programs below. The language in which Σ-specifications are written down is called the *Σ-language.* Thus within the concept of Σ-programming we distinguish the Σ-language, in which specifications for problems are written, and, in particular, Σ-specifications, and Σ-machines—an abstract construction, "fulfilling" a Σ-specification. In this paper we will basically concern ourselves with Σ-languages and Σ-specifications. Representations by a Σ-machine will have a very general, schematic character, since we assume that "Σ-machines" as a conceptual notion should be formulated after such notions as "Σ-specification" and "Σ-program" are made precise and developed. However, at the same time we mention that, like the initial text, the "executable" codes of Σ-programs will be by construction in

([1])Naturally, in the present case we are speaking not of a real computing device, but of some abstract notion.

one and the same language: the Σ-language. Only the level of abstraction will distinguish them.

Since we have brought up levels of abstraction, it is appropriate to remark that the process of Σ-programming begins with a description of the most abstract (relative to the Σ-machine level) specification of the problem to be solved, and ends with the creation of a completely concrete (i.e., acceptable by a Σ-machine) Σ-program. Since the initial specification can be essentially different in form from a Σ-specification, the problem arises of constructing a Σ-program corresponding to a specification of a problem. The process of construction of a Σ-program from the initial specification (including the construction of the initial specification itself) will be called *Σ-design*. We will deal in later publications with the methodological principles which underlie Σ-design. Here we dwell in more detail on the methodological attitudes which constitute the basis of the Σ-language.

We have already spoken above about the fact that both specifications for problems and Σ-specifications (including Σ-programs) are syntactical constructs in the Σ-language. Since dynamic aspects are closely interwoven with static ones in real mathematical and programming activities for solving problems, there must be methods in the Σ-language which adequately reflect both aspects. Thus the Σ-language, in whose terms all problems will be formulated, and their solutions described in the form of Σ-programs, must allow us to reflect completely and in a convenient manner specific features of the problem areas under investigation, their dynamic and static aspects. With this aim, the Σ-language will contain an extended *system of commands* (initial operations and methods of composing new ones), and a *system of declarations*. The role of the latter will be played by formulas in a special language of logic with first-order predicates. Since the Σ-specification should be "fulfillable" descriptions, the requirement follows that they should have "natural" command interpretations. This circumstance also explains why we take as Σ-specifications formulas of a special form: the so-called *Σ-formulas* (cf. §3). The "fulfillment" of Σ-formulas will be included in the verification of their *truth* in models. Such a verification can be carried out by various means, including syntactical ones (derivability). Here the main thing is the effectivity of the methods (their effective computability). If the user of the Σ-language is interested only in the possibility of specifying problems in the form of Σ-specifications, then the choice of strategies for their fulfillment lies with the Σ-translator. Thus we are concerned with universal "built-in" strategies for verification of the truth of Σ-formulas. If the user is interested in the effectivity of Σ-specifications, then the Σ-language should permit him to describe the "computation" strategy, taking such a description as one of the components of Σ-programs. Thus in general a Σ-program is presented as a construction consisting of two components: the first component talks about *what* the Σ-program is supposed to do (the logical component) and the second describes *how* this "what" is supposed to be done. Here both components have a particular declarative form: the form of Σ-formulas. Such a view of a Σ-program gives the Σ-design process an essentially mathematical investigative character: it is necessary not only to construct the Σ-specification, but also to transform it into a Σ-program, i.e., to carry out the process of verification of truth effectively.

Another way to improve the effectivity of a Σ-program is its *transformation.* It is natural to require that such a transformation should be carried out in the framework and with the methods of the Σ-language. Moreover, we require that the possibility of logically establishing the correctness of a transformation of Σ-programs should be supplied within the framework of the Σ-language. Here we have touched on a circumstance which we would like to dwell on in greater detail. It is a matter of the *problem of self-applicability.* In the present case this problem interests us not only in the sense in which it is understood in the mathematical theory of computation (for example, when the meaningfulness of functional expressions of the form $f(f)$ is studied), but as in the study of a language, where the description of constructions in it may be carried out in the framework and with the methods of the very same language. Here, as opposed to, say, Peano arithmetic, where the hierarchy of descriptions of the objects constructed and the constructions themselves can be embedded in the initial formalism by means of suitable numberings, which themselves are arithmetical constructions, we need to speak of the *clear* description and representation of all necessary constructions, as for the user the requirements of simplicity, convenience, and comprehensibility are essential. Such "self-applicable" languages (as is, for example, a natural language) having, on one hand, the virtues described above, have semantics which are extremely difficult to study. Therefore, in requiring "clear self-applicability" of the Σ-language, we are justified in expecting that the construction of such a language and its theory will be nontrivial, and also that the architecture of the solutions in the Σ-machines we create will be nontrivial.

When creating a programming language it is necessary above all to have a good concept of its semantics, and, in particular, to solve the problem of the semantical representation of the concept "constructive object". By a *constructive object* we will understand a finite object supplied with an "internal system of coordinates" (together with the choice of some initial element, the "initial coordinate"), which makes it possible to distinguish each of the components of the object in a well-defined way. As examples of constructive objects we may cite finite words, graphs, trees, automata, etc. For us the most adequate semantical representation of the concept "constructive object" will be a *finite list*: on one hand, finite lists have a definite universality for encoding: all the examples of constructive objects recalled above allow natural representations in the form of lists; and on the other hand, structures given as "lists" allow a satisfactory machine realization.

A constructive object may be regarded as an element of some "real world", given formally, say, in the form of a many-sorted model (possibly with some "superstructure" on the model). Since we are concerned with the constructivity of the representation, the most natural such model is thought of *intensionally*: either as a procedure, generating elements of the model (and elements of the "superstructure"), or as a constructive model (positive or strongly constructive). Such models are naturally called *data structures*, or just *data*, even if they are of the most abstract nature, and their elements may be called *data elements*. Furthermore, a constructive object may be considered also as an object of the "world" represented in the form of a *system of notations* about this world, for example in the form of a *formal theory*. Such a theory distinguishes, among all data, those which are models of it. In a definite sense the theory acts as one description of the whole class of its distinguished models, i.e., as the *type* of a class of data structures. For just this reason we will consider formal theories as the most adequate formalization of the concept of an "abstract data type".

In solving a problem we will be obliged in one way or another, besides using knowledge on some model representing the "world" to which the problem refers, to manufacture additional auxiliary constructions over the model, manipulating them and reasoning about them. Since we are interested in the solution of problems on computers, i.e., constructive algorithmic solutions, it is natural that such auxiliary constructs of necessity should be constructive objects and be of a transparent character, as already mentioned. Thus we need a model for which a whole family of constructive objects which we require for the solution of a problem may be simply and conveniently realized. Therefore, in speaking of a model above as a formal representation of a problem area, we recalled also a certain "superstructure" on it. List structures on many-sorted models are such superstructures. A model and its superstructure will also be the formal construct in which subject areas of interest to us will be represented, as also all the means of manipulation by the constructions we need. Working in such a "two-layer" model, we will use syntactic descriptions in terms of the Σ-language of constructions. Therefore, it is completely natural to require that for the "reasonable" descriptions in our model there should be constructions (realizations of the descriptions). Here the structuralization of a list superstructure should find its natural expression in the language working with the model.

This list superstructure as the superstructure of all finite lists on the initial many-sorted model will be constructed in a sufficiently uniform way. Therefore, we can regard it as a model (data structure) for some formal theory. Thus, working within the framework of the Σ-language, we will deal with formal theories where one part is the theory of the subject area, and the other is the formal theory of lists over the given area. The study of such formal theories will be carried out in the framework of the so-called list expansion GES (cf. §3). The interesting mathematical theory of list expansions will also be our tool later on in the construction of the theory of the Σ-language as well as the theory of Σ-machines. In the framework of this theory we will also study the logico-mathematical models of systems for the design of Σ-programs. In other words the theory GES, exposed in the following paragraphs, is the theoretical basis of Σ-programming.

As we will see later on, in the theory GES there are very powerful methods for the representation and investigation of a broad class of effective constructions, usable in Σ-programming. The collection of all constructions turns out to be closed with respect to various recursive operations, which also provide the possibility of constructing and using the various formal theories in the Σ-language.

The core of the Σ-language will be a set of initial commands, which may be regarded as the simplest operations of the language, and a family of initial predicates. In terms of this basis it will be possible to construct even necessarily complex Σ-programs. However, as will be shown in the theory GES, in the Σ-language there is a possibility of a natural conservative extension of the basic concepts. The necessity of such an extension has many causes. The main one is the convenience of the user of the Σ-language.

The basic mechanism for a conservative extension of the signature of the Σ-language will be the construction of individual objects for systems of quasi-identity, Σ-definability, and representability. Such a mechanism as "extraction of programs from constructive proofs" appears as the most promising (however, this requires additional investigation) for the conservative extension of the basis of the Σ-language. It is proposed to use this same mechanism also for the design of Σ-programs. Such a mechanism of constructive extension will play the main role in the question of the problem orientation of the Σ-language, whose solution will be included in the generation of abstract data types formally representing concepts of the subject area under consideration. Again, in view of the "self-applicability" property of the Σ-language, the given mechanisms and constructions can be used and improved in the framework of the language and its methods.

It is fundamental to assume (but again this requires additional investigations) that in the Σ-language one can construct (define) commands of higher levels of abstraction (functionals of finite type) which would make it possible to construct functional programs. This means that Σ-programming is indeed a concept which synthesizes within itself a large portion of the logico-mathematical approaches to programming: logical, functional, and transformational programming, the concept of "abstract data type", program synthesis, etc. Each of these approaches can be expressed in Σ-programming.

§3. The theory GES of list expansions

Let $\mathfrak{M}$ be a many-sorted model of signature σ_0, I the set of sorts of this model, and $\{M_i\}_{i\in I}$ the family of its basic sets. Below we will consider models with everywhere defined operations and predicates, although it would be good to investigate a more general case as well: partial models, in which it is necessary to distinguish a sort ω from the other sorts of the model, with $M_\omega = \{\perp\}$, where $\perp$ is interpreted as "undefined". The elements of the model $\mathfrak{M}$ will be considered as *atoms*, from which lists are constructed, then lists of lists, etc. The list superstructure over the model is denoted by $S(\mathfrak{M})$. The empty list nil occupies a special place in it, which may also be considered at the same time as an atom. We may regard $S(\mathfrak{M})$ as a new sort $\mathfrak{s}$ of objects, intended to be used to define various constructions on $\mathfrak{M}$, and to study their algorithmic nature. For this, besides the elements of the set $S(\mathfrak{M})$ themselves, it is necessary to introduce additional operations and relations. We have already introduced the 0-ary operation nil. Furthermore, we need the following operations and relations:([2])

head: $S(\mathfrak{M}) \to S(\mathfrak{M}) \times (\bigcup_{i\in I} M_i)$ is the operation of taking the last element in the list, if the list is nonempty, and giving nil if the list is empty.

tail: $S(\mathfrak{M}) \to S(\mathfrak{M})$ is the operation of taking the remainder of a list, from which the last element is deleted, if the list is nonempty, and giving nil otherwise.

cons: $S(\mathfrak{M}) \times (S(\mathfrak{M}) \cup (\bigcup_{i\in I} M_i)) \to S(\mathfrak{M})$ is the operation of adjunction to a list of a new element as the last one.

conc: $S(\mathfrak{M}) \times S(\mathfrak{M}) \to S(\mathfrak{M})$ is the operation of concatenation of two lists

$\in\, \subseteq (S(\mathfrak{M}) \cup (\bigcup_{i\in I} M_i)) \times S(\mathfrak{M})$ is the relation "is an element of the list".

$\sqsubseteq\, \subseteq S(\mathfrak{M}) \times S(\mathfrak{M})$ is the relation "is an initial segment of".

$S^1 \subseteq S(\mathfrak{M}) \cup (\bigcup_{i\in I} M_i)$ is the unary relation "is a list".

We consider the standard list superstructure $S^{\text{fin}}(\mathfrak{M})$, which presents the main interest for Σ-programming and which consists of all finite lists. We set:

$S^0(\mathfrak{M}) =$: {finite words (linear lists) of elements of $M = \bigcup_{i\in I} M_i$};

$S^1(\mathfrak{M}) =$: {linear lists of elements of $S^0(\mathfrak{M}) \cup \mathfrak{M}$};

$\dots$;

$S^{n+1}(\mathfrak{M}) =$: {linear lists of elements of $S^n(\mathfrak{M}) \cup \mathfrak{M}$}; etc.

Let $S^{\text{fin}}(\mathfrak{M}) = \bigcup_{n\geq 0} S^n(\mathfrak{M})$.

If the list α consists of the elements $\alpha_1, \dots, \alpha_n$ in a given sequence, then we will write $\alpha = \langle \alpha_1, \dots, \alpha_n \rangle$. We define the operations head, tail, cons, conc,

([2])The operations head and tail are somewhat different from the usual ones. Our choice of semantics for these operations was dictated only by considerations of convenience.

and $\underline{\text{nil}}$ on $S^{\text{fin}}(\mathfrak{M})$, setting:

$\underline{\text{head}}\,(\langle\alpha_1,\dots,\alpha_{n+1}\rangle) = \alpha_{n+1}$ and $\underline{\text{head}}\,(\underline{\text{nil}}) = \underline{\text{nil}}$;

$\underline{\text{tail}}(\langle\alpha_1,\dots,\alpha_n,\alpha_{n+1}\rangle) = \langle\alpha_1,\dots,\alpha_n\rangle$ and $\underline{\text{tail}}(\langle\alpha_1\rangle) = \underline{\text{tail}}\,(\underline{\text{nil}}) = \underline{\text{nil}}$;

$\underline{\text{cons}}\,(\langle\alpha_1,\dots,\alpha_n\rangle,\beta) = \langle\alpha_1,\dots,\alpha_n,\beta\rangle$;

$\underline{\text{conc}}\,(\langle\alpha_1,\dots,\alpha_n\rangle,\langle\beta_1,\dots,\beta_m\rangle) = \langle\alpha_1,\dots,\alpha_n,\beta_1,\dots,\beta_m\rangle$;

$\underline{\text{nil}} = \langle\ \rangle$;

$\beta \in \langle\alpha_1,\dots,\alpha_n\rangle \Leftrightarrow [\beta = \alpha_i$ for some $i,\ 1 \le i \le n]$;

$\langle\alpha_1,\dots,\alpha_n\rangle \sqsubseteq \langle\beta_1,\dots,\beta_m\rangle \Leftrightarrow [n \le m$ and $\alpha_i = \beta_i$ for some $i,\ 1 \le i \le n]$;

$\alpha \in S^1 \Leftrightarrow \alpha \in S^{\text{fin}}(\mathfrak{M})$.

We denote by HFS($\mathfrak{M}$) the model obtained from $\mathfrak{M}$ by adding a new sort s of elements of $S(\mathfrak{M})$ with the operations discussed above. We will call it the *superstructure of hereditarily finite lists over* $\mathfrak{M}$.

We now define a formal language of computation with first-order predicates for working with this model. We denote by σ the signature $\sigma_0 \cup \{\underline{\text{head}}, \underline{\text{tail}}, \underline{\text{cons}}, \underline{\text{conc}}, \underline{\text{nil}}, \in, \sqsubseteq, S^1\}$ and we put $\sigma^* =: \sigma \cup \{Q_1^{m_1},\dots,Q_n^{m_n},\dots\}$, where $Q_i^{m_i}$ $(i > 1)$ is an m_i-ary predicate symbol not occurring in σ_0. We remark that if I is the set of sorts of the signature σ_0, then $I \cup \{s\}$ is the set of sorts of the signatures σ and σ^*. Besides these sorts, we will require one more construction: *type*, by which we will understand a finite ordering of a set $\langle K_1,\dots,K_n\rangle$ of subsets of I. Below we will assume that each symbol of the signature lies in a definite type.

Terms (of the signature σ_0, σ, or σ^*) are defined in the usual way.

Variables and constants of any types of the form $\langle K\rangle$, $K \subseteq I \cup \{s\}$ will be terms of the same types.

If F is an operation symbol of the type $\langle K_1,\dots,K_n,K\rangle$ (thus F is interpreted in the model $\mathfrak{M}$ as an operation:

$$\overline{F}\colon \left(\bigcup_{i\in K_1} M_i\right) \times\cdots\times \left(\bigcup_{i\in K_n} M_i\right) \to \bigcup_{i\in K} M_i\Bigg),$$

and $t_1,\dots,t_n$ are terms suitable for F (i.e., of the necessary type), then $F(t_1,\dots,t_n)$ is the term of type $\langle K\rangle$.

There are no other terms.

Formulas (of the signature σ_0, σ, or σ^* (3) are defined in an equally standard way.

If t_1 and t_2 are terms of the same type, then "$t_1 = t_2$" is a formula.

If P is an n-place predicate symbol of the type $\langle K_1,\dots,K_n\rangle$, and $t_1,\dots,t_n$ are terms suitable for P, then $P(t_1,\dots,t_n)$ is a formula.

If φ and ψ are formulas, then $(\varphi \vee \psi)$, $(\varphi \mathbin{\&} \psi)$, $(\varphi \to \psi)$, $\neg\varphi$, $\forall x\varphi$, and $\exists x\varphi$ are formulas, where x is a variable of any type $\langle K\rangle$.

(3) An exception is the last clause of the definition of formulas.

If φ is a formula, t is a term of type $\langle\{s\}\rangle$, and x is a variable of the type $\langle I \cup \{s\}\rangle$, then the expressions $(\forall x \in t)\varphi$, $(\exists x \in t)\varphi$, $(\forall x \subseteq t)\varphi$, and $(\exists x \subseteq t)\varphi$ will also be formulas; the quantifiers in these expressions are called *bounded*.

DEFINITION 1. a. A formula in which only bounded quantifiers are encountered is called a *formula with bounded quantifiers*, or a Δ_0-*formula*.

b. Any formula belonging to the smallest class of formulas containing all Δ_0-formulas, and closed with respect to disjunctions and conjunctions of formulas, as well as prefixing existential quantifiers, is called a Σ-*formula*. For Σ-formulas in the signature σ^* it is also necessary to require that the predicate symbols $Q_1^{m_1}, \dots, Q_n^{m_n}, \dots$ occur only positively in such formulas.

We call a list of two elements α_1, α_2 occurring in α and in the specified order an *ordered pair*, and we denote it $\langle\alpha_1, \alpha_2\rangle$. We remark that the property "to be an ordered pair" is expressible in the form of a Δ_0-formula:

$$\mathrm{ORDPAIR}(\alpha) = (\exists\alpha_1 \in \alpha)(\exists\alpha_2 \in \alpha)(\alpha = \underline{\mathrm{cons}}(\underline{\mathrm{cons}}(\underline{\mathrm{nil}}, \alpha_1), \alpha_2)).$$

Therefore, the notation $\langle\alpha_1, \alpha_2\rangle$ is in fact an abbreviation for $\underline{\mathrm{cons}}(\underline{\mathrm{cons}}(\underline{\mathrm{nil}}, \alpha_1), \alpha_2)$.

The list α will be called a *list function of the list* β if the elements of α are ordered pairs in which the first elements are initial segments of the list β. Here the elements of the list α are ordered in the order of increasing first elements of the pairs from α (relative to the predicate $\sqsubseteq$). Thus the predicate $\underline{\mathrm{LF}}(\alpha, \beta)$ ("α is a list function of β") may be represented by the following Δ_0-formula:

$$\begin{aligned}\underline{\mathrm{LF}}(\alpha, \beta) =: (\forall\gamma \in \alpha)(&\mathrm{ORDPAIR}(\gamma)\ \&\ (\forall\alpha_1 \in \gamma)(\forall\alpha_2 \in \gamma) \\ &\times (\langle\alpha_1, \alpha_2\rangle = \gamma \to \alpha_1 \sqsubseteq \beta)) \\ \&(\forall\delta \sqsubseteq \beta)&(\exists\gamma \in \alpha)(\underline{\mathrm{cons}}(\underline{\mathrm{nil}}, \delta) = \underline{\mathrm{tail}}(\gamma) \\ \&(\forall\alpha' \sqsubseteq \alpha)&(\forall\alpha'' \sqsubseteq \alpha')(\exists\delta' \in \underline{\mathrm{tail}}\ (\underline{\mathrm{head}}(\alpha'))) \\ &\times (\exists\delta'' \in \underline{\mathrm{tail}}\ (\underline{\mathrm{head}}(\alpha'')))(\delta'' \sqsubseteq \delta').\end{aligned}$$

A list α which is a list function of the list β may be interpreted as a tabulated function whose arguments are initial portions of the list β. Here $\sqsubset =: \sqsubseteq \setminus =$. We call the list α a *hereditary list function of* β (corresponding to the predicate $\underline{\mathrm{HLF}}(\alpha, \beta)$) if α is a list function of β and for all $\gamma \in \alpha$ we have

1) if $\underline{\mathrm{head}}\ (\underline{\mathrm{tail}}(\gamma)) = \underline{\mathrm{nil}}$, then $\underline{\mathrm{head}}(\gamma) = \underline{\mathrm{nil}}$;

2) if $\gamma = \langle\alpha_1, \alpha_2\rangle$ and $\alpha_1 = \underline{\mathrm{cons}}(\alpha_1', a)$, then $\alpha_2 = \underline{\mathrm{cons}}\ (\alpha_2', b)$ and $\langle\alpha_1', \alpha_2'\rangle \in \alpha$. $\langle\alpha_1', \alpha_2\rangle \in \alpha\ \&\ \alpha_2' = \alpha_2$.

We remark that the pair $\langle\alpha_1', \alpha_2'\rangle$ in the list α must occur directly after the pair $\langle\alpha_1, \alpha_2\rangle$.

Hereditary list functions α of the list β may be regarded as "regular constructions" of tabulated functions, following the previous notation.

The formal theory GES of list superstructures for the model $\mathfrak{M}$ has as its axioms the universal closures of the following formulas.

AXIOM OF THE EMPTY LIST. $\neg(\exists\delta)(\delta \in \underline{\text{nil}})$.

AXIOM OF UNIQUENESS. $(\underline{\text{cons}}(\alpha, \delta) = \underline{\text{cons}}(\alpha', \delta') \to (\alpha = \alpha' \ \&\ \delta = \delta'))$, where α and α' are variables of type $\langle\{s\}\rangle$ (i.e., the type of *lists*), and δ and δ' are of the type $\langle I \cup \{s\}\rangle$.

Below, when the context uniquely determines the types of the variables under consideration we will omit the description of the types of these variables.

AXIOMS OF LIST OPERATIONS.

$\underline{\text{tail}}\ (\underline{\text{cons}}(\alpha, \delta)) = \alpha$;

$\underline{\text{head}}\ (\underline{\text{cons}}(\alpha, \delta)) = \delta$;

$(\neg(\alpha = \underline{\text{nil}}) \to \underline{\text{cons}}\ (\underline{\text{tail}}(\alpha),\ \underline{\text{head}}(\alpha)) = \alpha)$;

$\underline{\text{tail}}\ (\underline{\text{nil}}) = \underline{\text{nil}}$;

$\underline{\text{head}}\ (\underline{\text{nil}}) = \underline{\text{nil}}$;

$\underline{\text{cons}}\ (\underline{\text{conc}}(\alpha, \beta), \gamma) = \underline{\text{conc}}(\alpha, \underline{\text{cons}}(\beta, \gamma))$;

$\underline{\text{conc}}\ (\underline{\text{conc}}(\alpha, \beta), \gamma) = \underline{\text{conc}}(\alpha, \underline{\text{conc}}(\beta, \gamma))$;

$\underline{\text{conc}}\ (\underline{\text{nil}}, \alpha) = \underline{\text{conc}}(\alpha, \underline{\text{nil}}) = \alpha$.

$(\alpha \in \underline{\text{cons}}(\beta, \delta) \Leftrightarrow (\alpha \in \beta \vee \alpha = \underline{\text{cons}}(\beta, \delta))) \ \&\ (\delta \in \underline{\text{cons}}(\alpha, \delta') \Leftrightarrow (\delta \in \alpha \vee \delta = \delta'))$.

AXIOMS OF INDUCTION.

$$([\Phi]^x_{\underline{\text{nil}}} \ \&\ (\forall\alpha)(\forall\delta)([\Phi]^x_\alpha \to [\Phi]^x_{\underline{\text{cons}}(\alpha,\delta)}) \to \alpha\Phi),$$

where Φ is an arbitrary formula and the notation $[\Phi]^x_t$ denotes the substitution of the term t in place of all free occurrences of the variable x in Φ, if this does not lead to clashes between variables.

AXIOM OF EQUALITY OF CAPACITY.

$$\alpha = \beta \Leftrightarrow (\forall\gamma \sqsubseteq \alpha)(\gamma \sqsubseteq \beta \ \&\ (\neg(\gamma = \beta) \to (\exists\delta \in \alpha)(\underline{\text{cons}}(\gamma, \delta) \sqsubseteq \alpha \ \&\ \underline{\text{cons}}(\gamma, \delta) \sqsubseteq \beta))).$$

AXIOM OF FOUNDATION.

$$(\forall\delta)((\forall\gamma \in \delta)\Phi(\gamma) \to \Phi(\delta)) \to \forall\alpha\Phi(\alpha).$$

AXIOM OF Δ_0-SEPARATION. If $\Phi(\alpha)$ is a Δ_0-formula with the free variable α, then

$$\begin{aligned}(\forall\beta)(\forall\alpha)(&\underline{\text{HLF}}(\alpha, \beta) \ \&\ (\forall\beta' \sqsubseteq \beta)(\forall\delta \in \beta)(\underline{\text{cons}}(\beta', \delta) \sqsubseteq \beta \\ &\to (\forall\gamma_1 \in \alpha)(\forall\gamma_2 \in \alpha)(\underline{\text{head}}(\underline{\text{tail}}(\gamma)) = \beta' \ \&\ \underline{\text{head}}(\underline{\text{tail}}(\gamma_2)) = \underline{\text{cons}}(\beta', \delta) \\ &\to (\Phi(\delta) \to \underline{\text{cons}}(\underline{\text{head}}(\gamma_1), \delta) = \underline{\text{head}}(\gamma_2)) \\ &\&\ (\neg\Phi(\delta) \to \underline{\text{head}}(\gamma_1) = \underline{\text{head}}(\gamma_2)))) \\ &\&\ (\forall\gamma \in \alpha)(\underline{\text{head}}(\underline{\text{tail}}(\gamma)) = \underline{\text{nil}} \to \underline{\text{head}}(\gamma) = \underline{\text{nil}})).\end{aligned}$$

This axiom expresses the existence for every list β of a tabulated function which distinguishes in it just those elements for which property Φ is satisfied.

AXIOM OF Δ_0-CHOICE. If $\Phi(x, y)$ is a Δ_0-formula, then

$$(\forall\beta)[(\forall\alpha \in \beta)(\exists\delta)\Phi(\alpha,\delta) \to \exists\alpha\ (\underline{\text{HLF}}(\alpha,\beta)\ \&\ (\forall\gamma \in \alpha) \times (((\underline{\text{head}}(\underline{\text{tail}}(\gamma)) = \underline{\text{nil}}) \to \underline{\text{head}}(\gamma) = \underline{\text{nil}}) \ \&\ (\neg(\underline{\text{head}}(\underline{\text{tail}}(\gamma)) = \underline{\text{nil}}) \to \Phi(\underline{\text{head}}(\underline{\text{head}}(\underline{\text{tail}}(\gamma))), \underline{\text{head}}(\underline{\text{head}}(\gamma))))))].$$

This axiom speaks about the possibility, in every concrete case (i.e., for every list β), of realizing the "functional relation" defined by the formula Φ, restricting only to a certain list of α, and not the whole universe.

With this the list of axioms describing the list superstructure over the model $\mathfrak{M}$ is complete. We would like to call the reader's attention to one circumstance. The form of the axioms is somewhat awkward, but if they are analyzed attentatively, then the reason for this awkwardness is clear: the axioms are *constructive* in nature, since they speak not only of the existence of necessary objects, but also of how they are constructed. Such constructivity is a consequence of the *determinability* of list constructions in opposition to, say, set-theoretic constructions.

We consider certain properties of the theory GES, illustrating their adequacy for the goals with which we were concerned in §2.

First of all we remark that in GES it turns out that the traditional operations on lists are definable, for example the construction of a "list of lists":

$$(\forall\beta)(\exists\alpha)(\underline{\text{HLF}}(\alpha,\beta)\ \&\ (\forall\gamma \sqsubseteq \beta)(\forall\delta \in \beta)(\underline{\text{cons}}(\gamma,\delta) \sqsubseteq \beta \to (\exists\gamma_1 \in \alpha)(\exists\gamma_2 \in \alpha)(\underline{\text{head}}(\underline{\text{tail}}(\gamma_1)) = \gamma\ \&\ \underline{\text{head}}(\underline{\text{tail}}(\gamma_2)) = \underline{\text{cons}}(\gamma,\delta) \ \&\ (S^1(\delta) \to \underline{\text{conc}}(\underline{\text{head}}(\gamma_1),\delta) = \underline{\text{head}}(\gamma_2)) \ \&\ (\neg S^1(\delta) \to \underline{\text{cons}}(\underline{\text{head}}(\gamma_1),\delta) = \underline{\text{head}}(\gamma_2))\ \&\ (\forall\gamma \in \alpha)(\underline{\text{head}}(\underline{\text{tail}}(\gamma)) = \underline{\text{nil}} \to \underline{\text{head}}(\gamma) = \underline{\text{nil}})).$$

Now let Φ be a formula and v a variable not encountered in this formula. We will write $\Phi_{\in}^{(v)}$ to denote the result of changing all occurrences of unbounded quantifiers in Φ into bounded ones: $\exists\alpha$ becomes $\exists\alpha \in v$ and $\forall\alpha$ becomes $\forall\alpha \in v$; and we will write $\Phi_{\sqsubseteq}^{(v)}$ for the result obtained by changes of the following form: $\exists\alpha$ becomes $\exists\alpha \sqsubseteq v$ and $\forall\alpha$ becomes $\forall\alpha \sqsubseteq v$. If in the formula Φ, changes of all unbounded quantifiers into bounded ones have occurred, without mentioning the type of change ($\in$ or $\sqsubseteq$), then we will denote the result of the changes by $\Phi^{(v)}$. We remark that $\Phi_{\in}^{(v)}$, $\Phi_{\sqsubseteq}^{(v)}$, and $\Phi^{(v)}$ are Δ_0-formulas. If Φ were also a Δ_0-formula then $\Phi = \Phi^{(v)} = \Phi_{\in}^{(v)} = \Phi_{\sqsubseteq}^{(v)}$.

THEOREM 1 (The Σ-reflection principle). *For all Σ-formulas Φ in the signature σ,*

$$\text{GES} \vdash \Phi \Leftrightarrow \exists v \Phi_{\in}^{(v)}.$$

Below, a formula of the form $\exists\alpha\Phi(\alpha)$, where Φ is a Δ_0-formula, will be called a Σ_1-*formula*. Thus Theorem 1 says that every Σ-formula is equivalent to a Σ_1-formula.

Above we distinguished the class of Σ-formulas. On the other hand, one may distinguish the class of Π-formulas: the smallest class of formulas containing the Δ_0-formulas and closed with respect to disjunctions, conjunctions, and prefixing of bounded quantifiers or unbounded universal quantifiers. Clearly the negations of Σ-formulas are logically equivalent to Π-formulas, and conversely. A formula which is logically equivalent simultaneously to a Σ-formula and a Π-formula is called a *Δ-formula*.

THEOREM 2 (The Δ-separation principle). *For any Σ-formulas Φ and Ψ,*

$$\begin{aligned}\mathrm{GES} \vdash (\forall\alpha \in \beta)(\Phi(\alpha) \leftrightarrow \neg\Psi(\alpha)) \\ \rightarrow \exists\alpha(\underline{\mathrm{HLF}}(\alpha, \beta) \;\&\; (\forall\gamma \in \alpha)[(\underline{\mathrm{head}}(\underline{\mathrm{tail}}(\gamma)) = \underline{\mathrm{nil}} \rightarrow \underline{\mathrm{head}}(\gamma) = \underline{\mathrm{nil}}) \\ \&\; (\neg(\underline{\mathrm{head}}(\underline{\mathrm{tail}}(\gamma)) = \underline{\mathrm{nil}}) \rightarrow (\Psi(\underline{\mathrm{head}}(\underline{\mathrm{head}}(\underline{\mathrm{tail}}(\gamma)))) \\ \rightarrow \underline{\mathrm{cons}}(\underline{\mathrm{cons}}(\underline{\mathrm{nil}}, \underline{\mathrm{tail}}(\underline{\mathrm{head}}(\underline{\mathrm{tail}}(\gamma)))), \underline{\mathrm{tail}}(\underline{\mathrm{head}}(\gamma))) \in \alpha))]).\end{aligned}$$

COROLLARY 1.

$$\begin{aligned}\mathrm{GES} \vdash (\forall\alpha \in \beta)(\Phi(\alpha) \leftrightarrow \neg\Psi(\alpha)) \\ \rightarrow (\exists\alpha)((\forall\delta \in \beta)(\Phi(\delta) \rightarrow \delta \in \alpha)) \;\&\; (\forall\gamma \in \beta)(\Psi(\gamma) \rightarrow \neg(\gamma \in \alpha)).\end{aligned}$$

THEOREM 3 (The principle of Σ-choice). *For any Σ-formula Φ,*

$$\begin{aligned}\mathrm{GES} \vdash (\forall\beta \sqsubseteq a)(\exists\gamma)\Phi(\beta, \gamma) \\ \rightarrow (\exists\alpha)(\exists b)(\underline{\mathrm{STRLF}}(\alpha, a, b) \;\&\; (\forall\delta \sqsubseteq a)(\neg(\delta = \underline{\mathrm{nil}}) \rightarrow \Phi(\underline{\mathrm{head}}(\delta), \alpha(\delta)))),\end{aligned}$$

where the predicate $\underline{\mathrm{STRLF}}(\alpha, a, b)$ ("the list α structurally transforms the list a into b") is defined by the formula

$$\begin{aligned}&\underline{\mathrm{STRLF}}(\alpha, a, b) \\ &=: [\underline{\mathrm{HLF}}(\alpha, a) \& (\forall\gamma \in \alpha)(\forall\gamma' \in \alpha)(((\underline{\mathrm{head}}(\underline{\mathrm{tail}}(\gamma)) = \underline{\mathrm{nil}}) \rightarrow (\underline{\mathrm{head}}(\gamma) = \underline{\mathrm{nil}})) \\ &\quad \& (\neg(\underline{\mathrm{head}}(\underline{\mathrm{tail}}(\gamma)) = \underline{\mathrm{nil}}) \;\&\; (\underline{\mathrm{head}}(\underline{\mathrm{tail}}(\gamma)) = \underline{\mathrm{tail}}(\underline{\mathrm{head}}(\underline{\mathrm{tail}}(\gamma')))) \\ &\quad \rightarrow \underline{\mathrm{tail}}(\underline{\mathrm{head}}(\underline{\mathrm{tail}}(\gamma'))) = \underline{\mathrm{head}}(\underline{\mathrm{tail}}(\gamma)))) \;\&\; \underline{\mathrm{head}}(\underline{\mathrm{head}}(\alpha)) = b].\end{aligned}$$

REMARK. In the definition of the predicate $\underline{\mathrm{STRLF}}(\alpha, a, b)$, to every initial section of a the list α "relates" a *unique* initial segment of b, where the "transformation" α is monotonic for the inclusion of lists "$\sqsubseteq$". Therefore we may introduce the functional notation $\alpha(\gamma)$ for all $\gamma \equiv \alpha$, where $\alpha(\gamma)$ denotes the initial segment of the list b corresponding to γ under the structural transformation. By $\underline{\alpha(\gamma)}$ we will denote the last element of the list $\alpha(\gamma)$. The expression $\underline{\alpha(\gamma)} = \delta$ is sufficiently simply defined by the formula

$$\underline{\alpha(\gamma)} = \delta =: (\exists s \in \gamma)(s = \underline{\mathrm{cons}}(\underline{\mathrm{cons}}(\underline{\mathrm{nil}}, \gamma), \delta)).$$

We may act similarly in the case of $\underline{\mathrm{LF}}(\alpha, a)$, introducing again here the notation $\alpha(\gamma)$ for the value corresponding to α on the section $\gamma \sqsubseteq \alpha$ and denoting by $\underline{\alpha(\gamma)}$ the last element in the list $\alpha(\gamma)$. Thus all occurrences of the expression $\underline{\alpha(\gamma)} = \delta$ in formulas may be replaced by Δ_0-formulas expressing this equality.

DEFINITION 2. Let $\Phi(\overline{x}, y)$ be a Σ-formula of the signature σ, for which we have GES $\vdash (\forall\overline{x})(\exists! y)\Phi(\overline{x}, y)$, where $\overline{x} = \langle x_1, \ldots, x_n\rangle$, and let F be an n-place function symbol, not in the signature σ. We set

$$(F)\colon F(\overline{x}) = y \leftrightarrow \Phi(\overline{x}, y)$$

and we will say that F is *Σ-definable* in GES.

Similarly we can introduce the concept of a Σ-definable predicate.

THEOREM 4. *For all formulas $\Phi(\overline{z}, F)$ of the signature $\delta \cup \{F\}$, there are formulas Φ_0 of the signature σ such that* GES $+ (F) \vdash \Phi(\overline{z}, F) \leftrightarrow \Phi_0(\overline{z})$. *If Φ is a Σ-formula, so is Φ_0. If Φ is a Δ_0-formula, then there will be Σ- and Π-formulas Φ_0 and Φ_1 correspondingly* (in the signature σ) so that

$$\text{GES} + (F) \vdash \Phi \leftrightarrow \Phi_0, \qquad \text{GES} + (F) \vdash \Phi \vdash \Phi_1,$$

and GES $+ (F)$ is a conservative extension of GES.

REMARK. A similar theorem may be proved also for predicates. These results, in particular the last clause, have extraordinarily important meaning for Σ-programming in connection with the remarks made in §2 with regard to the definable constructs in the Σ-language, conservatively extending it.

The construction $\underline{\text{TC}}(\alpha)$—the transitive closure of the list α—also plays an exceptionally important role in the theory GES and in Σ-programming as a whole. Its definition is very awkward, and therefore we do not go through it here. We will say only that in terms of its meaning it is similar to the notion of transitive closure for sets. TC(α) as a Σ-definable function has great significance in the theory of Σ-recursion, allowing one to construct for any Σ-definable function a new recursively Σ-definable function. The possibility and correctness of such a construction is established in Theorem 5, which may be regarded as the analog of Gandy's theorem in the theory of admissible sets [1], [2]. Before passing to its formulation, we define a few constructions.

Let $\mathfrak{M}$ be a model of signature σ_0. The many-sorted model $\mathfrak{A}(\mathfrak{M})$ of signature σ, obtained from $\mathfrak{M}$ by taking the union of it with objects of the new sort s, will be called a *list extension* of $\mathfrak{M}$ if $\mathfrak{A}(\mathfrak{M}) \vdash$ GES. If $\mathfrak{A}(\mathfrak{M})$ is a list extension of $\mathfrak{M}$, then the set of all of the objects of sort s will be denoted by $S(\mathfrak{M})$ and called the *lists over* $\mathfrak{M}$, and the set $S(\mathfrak{M})$ itself will be the *list superstructure over* $\mathfrak{M}$. Below, the list extension $\mathfrak{A}(\mathfrak{M})$ will often be written as a pair $\langle\mathfrak{M}; S(\mathfrak{M})\rangle$. As an example of a list extension we may consider HFS$(\mathfrak{M}) =: \langle\mathfrak{M}; S^{\text{fin}}(\mathfrak{M})\rangle$.

Let $\Phi(\overline{x}, \overline{y}, Q)$ (where $\overline{x}$ and $\overline{y}$ are sequences of free variables) be Σ-formulas in the signature σ^*, having only positive occurrences of one predicate variable $Q \notin \sigma$, and let $\mathfrak{A}(\mathfrak{M})$ be a list extension of the model $\mathfrak{M}$. We define the operator

$$\Gamma_{\Phi,\overline{y}}(\widetilde{Q}) =: \{\overline{a} | \mathfrak{A}(\mathfrak{M}) = \Phi(\overline{a}, \overline{y}, Q)\}.$$

In view of the positivity of the occurrences of the variable Q in Φ the operator $\Gamma_{\Phi,\overline{y}}$ is monotone, and accordingly has a least fixed point.

THEOREM 5. *For each Σ-formula $\Phi(\overline{x}, \overline{y}, Q)$ of the signature σ^* having positive occurrences of the predicate variable Q, there is a Σ-formula $\Phi(\overline{x}, \overline{y})$ defining the least fixed point of the operator $\Gamma_{\Phi,\overline{y}}$ for any fixed values of the parameters.*

The theory of Σ-recursion can be developed in different directions. For example, similarly to Ershov [3], we can extend first-order logic on lists, defining dynamic logic over list extensions, which enjoys the same good properties as dynamic logic on admissible models. It might prove useful to have infinitary variants of Σ-formulas. Again in GES the question of the concept of abstract data types and the problem of the representation of one type of data in another may be solved as naturally as in [3]. In other words, there remains serious work to be done on the development of the theory GES.

§4. Σ-programming

In this section fundamental attention will be paid to the concept of Σ-programming. Earlier we described how in the general case Σ-programming consists of two parts: logical and directive. Here it will be a matter only of the logical component of a Σ-program. The concept of Σ-programming itself will be reformulated not as a syntactical category, but at an interesting mathematical level. Thus the syntactic aspect of the Σ-language remains outside the scope of this article. Later a separate article will be devoted to it.

Let σ_0 and σ be the signatures defined in §3. We denote by σ^* the enrichment of the signature σ by predicate and function variables. We will say that the functional variable F *occurs in the formula* Φ *positively* if all the elementary formulas in which F is found occur positively.

We consider a Σ-formula $\Phi(\overline{x}; Q, F)$ where $\overline{x}$ is a sequence of all the free variables, of length n, and Q and F are sequences of predicate and function variables occurring positively in Φ. Let Q be an n-place predicate variable, and F an $(n-1)$-place function symbol of $\sigma^*\backslash\sigma$.

DEFINITION 3. Expressions of the form

$$\underline{\text{def}}\, Q(\overline{x})\colon \Phi(\overline{x}, Q, F), \tag{1}$$

and of the form

$$\underline{\text{def}}\, F(x_1, \dots, x_{n-1}) = x_n\colon \Phi(\overline{x}, Q, F) \tag{2}$$

(under the condition that in the last case the formula Φ defines a functional equality) are called *Σ-definitions* (of the signature σ^*). If the sequence $\overline{x}$ is empty, then expressions of the form (1) are also admissible, and the corresponding Σ-formulas are called *facts*.

EXAMPLE 1.(4)

a) $\underline{\text{def}}\, \underline{\text{FACT}}(x, y)\colon [\exists z((x = 0 \to y = 1) \vee (\neg(x = 0)$
$\to [(y = x * z)\ \&\ \underline{\text{FACT}}(x - 1, z)])]$;

(4)Below we will take as the signature σ_0 the signature of elementary arithmetic.

b) $\underline{\text{def}}\, x! = y: ((x = 0 \rightarrow \overline{y} = 1) \vee (\neg(x = 0) \rightarrow y = x * (x - 1)!))$.

In this example both Σ-definitions represent the same predicate. But where in case a) we used a Σ-formula as the definition, in case b) the definition is even a Δ_0-formula.

Below, as in Example 1, we will often encounter constructions of the type $(\Phi \,\&\, \Psi_1) \vee (\neg\Phi \,\&\, \Psi_2)$, where Φ, Ψ_1, and Ψ_2 are Δ-formulas. We will denote this in more usual programming terminology by: $\underline{\text{if}}$ Φ $\underline{\text{then}}$ Ψ_1 $\underline{\text{else}}$ Ψ_2.

In Σ-definitions we will distinguish the left part of the definition, standing before the colon, which will be called the *header*, and the right side, called the *body* of the Σ-definition. If the Σ-definition is a fact, then its header will also be called the *indicator*.

EXAMPLE 2. $\underline{\text{def}}$ $\underline{\text{TRUE}}$: $0 = 0$.

We now consider a finite set D of Σ-definitions. We will call it *regular*, if

a. among the headers of its Σ-definitions we do not encounter expressions with unary predicates and/or function symbols; and

b. the sequence of free variables in the Σ-definitions are pairwise disjoint.

Let $S_1(\overline{x}_1), \ldots, S_n(\overline{x}_n)$ be the sequence of headers corresponding to Σ-definitions from the set D, and $\underline{\text{Q}}$ and $\underline{\text{F}}$ the sequences (possibly empty) of predicate and function symbols occurring in the body of the Σ-definitions in D, but not found in the left sides of the Σ-definitions.

DEFINITION 4. An expression of the form

$$\langle\text{identifier}\rangle(S_1(\overline{x}_1), \ldots, S_n(\overline{x}_n); \mathbf{Q}; \mathbf{F}) \colon D \;\underline{\text{end}}, \tag{3}$$

where D is a regular family of Σ-definitions, and $S_i(\overline{x}_i)$ for $i = 1, \ldots, n$, $\underline{\text{Q}}$, and $\underline{\text{F}}$ were defined above, is called a *Σ-scheme* (of the signature σ^*).

In (3), ⟨identifier⟩ is called the *name* of the scheme, $S_i(\overline{x}_i)$ for $i = 1, \ldots, n$ are called the *inputs* of the scheme, the $\overline{x}_j$, $j = 1, \ldots, n$, are the *formal object parameters* of the jth input, the elements of $\underline{\text{Q}}$ or $\underline{\text{F}}$ are the *predicate* (respectively, *functional*) *formal parameters.*

A Σ-scheme is *closed* if the list of its predicate and function formal parameters is empty. Otherwise the Σ-scheme is called *open.*

EXAMPLE 3 (a closed Σ-scheme).

$$
\begin{aligned}
&\text{HOD}\ (\text{GCD}(x, y, z),\ \text{MIN}(x_1, y_1) = z_1,\ \text{DIV}(x_2, y_2), x_3 \le y_3, x_4 > y_4) :\\
&\underline{\text{def}}\ \text{GCD}(x, y, z) \colon (x > 0 \;\&\; y > 0 \;\&\; \underline{\text{DIV}}(z, x) \;\&\; \underline{\text{DIV}}(z, y)\\
&\qquad\qquad \& \; (\forall w \le \underline{\text{MIN}}(x, y))(\underline{\text{DIV}}(w, x) \;\&\; \underline{\text{DIV}}(w, y) \rightarrow w \le z));\\
&\underline{\text{def}}\ \underline{\text{MIN}}(x_1, y_1) = z_1 \colon \underline{\text{if}}\ x_1 \le y_1\ \underline{\text{then}}\ z_1 = x_1\ \underline{\text{else}}\ z_1 = y_1;\\
&\underline{\text{def}}\ \underline{\text{DIV}}(x_2, y_2) \colon (x_2 > 0 \;\&\; y_2 > 0 \;\&\; \exists z(z * x_2 = y_2));\\
&\underline{\text{def}}\ x_3 \le y_3 \colon \exists z(x_3 + z = y_3);\\
&\underline{\text{def}}\ x_4 > y_4 \colon [(y_4 \le x_4) \;\&\; \neg(x_4 = y_4)];\\
&\qquad\qquad\qquad\qquad\qquad \underline{\text{end}}.
\end{aligned}
$$

We note that in the definition of MIN it would be possible to use a still more indeterminate definition:

$$\underline{\text{def}}\ \underline{\text{MIN}}(x_1, y_1) = z_1\colon ((x_1 \le y_1 \to z_1 = x_1) \vee (y_1 \le x_1 \to z_1 = y_1)).$$

The following remark concerning Example 3 is connected with the fact that the definition of the greatest common divisor may be put in the form of a single Σ-definition. But first we need to demonstrate the idea of a structurization of knowledge in a "world" and the role of Σ-schemes in this structurization; and secondly, a Σ-scheme with many inputs, as we shall see later, is a richer, multipurpose structure in comparison with a single-input scheme.

From Theorem 5 we obtain the following result, which has a direct relation to the execution of Σ-programs on Σ-machines.

THEOREM 6. *Let J be a closed Σ-scheme. Then J defines a monotone operator Γ_J, and the least fixed point of this monotone operator is a Σ-definable set.*

Let $J(S_i(\overline{x}_i); \cdots ; S_n(\overline{x}_n); \cdots)$ be some Σ-scheme with inputs $S_i(\overline{x}_i)$, $i = 1, \ldots, n$. An expression of the form $[\overline{t} =: x]J.S_i$, where $\overline{t}$ is a sequence of terms of the signature σ^*, in which the number of elements is no less than the number of elements of the sequence $\overline{x}_i$, is called a *call* of the Σ-scheme J at the point S_i with actual parameters $\overline{t}$. Further, we allow the occurrence in Σ-formulas of calls on Σ-schemes, considering the latter as atomic formulas, with the proviso that they occur positively in the Σ-formula.

EXAMPLE 4.

$\underline{\text{ORD}}\ (\underline{\text{ORDLIST}}(\alpha), \underline{\text{FALSE}}, \underline{\text{TRUE}})$:
$\underline{\text{def}}\ \underline{\text{ORDLIST}}(\alpha)$: $\underline{\text{if}}\ \neg S^1(\alpha)\ \underline{\text{then FALSE}}$
$\underline{\text{else}}\ [(\forall x \in \alpha)[\neg S^1(x)\ \&\ \underline{\text{if}}\ (\underline{\text{tail}}(\alpha) = \underline{\text{nil}})\ \underline{\text{then TRUE}}$
$\quad\quad \underline{\text{else}}\ (\forall \beta \sqsubseteq \alpha)(\underline{\text{cons}}(\beta, x) \in \alpha \to [x, \underline{\text{head}}(\beta),$
$\quad\quad\quad \underline{\text{head}}(\beta) =: x_1, y_1, z_1]\underline{\text{HOD}}.\underline{\text{MIN}})$;
$\underline{\text{def}}\ \underline{\text{FALSE}}: 0 = 1$;
$\underline{\text{def}}\ \underline{\text{TRUE}}: 0 = 0$;

$\underline{\text{end}}$.

We remark that it is possible to introduce Σ-schemes recursively, defining the predicate ORDLIST:

$$\begin{aligned}
&\underline{\mathrm{ORD}}1\ (\underline{\mathrm{ORDLIST}}1(\alpha),\ \underline{\mathrm{FALSE}},\ \underline{\mathrm{TRUE}}):\\
&\quad \underline{\mathrm{def}}\ \underline{\mathrm{ORDLIST}}1(\alpha):\ \underline{\mathrm{if}}\ \neg S^1(\alpha)\ \underline{\mathrm{then\ FALSE}}\\
&\qquad \underline{\mathrm{else}}[(\forall x \in \alpha)[\neg S^1(x)\ \&\ \underline{\mathrm{if}}\ (\underline{\mathrm{tail}}(\alpha) = \underline{\mathrm{nil}})\\
&\qquad\quad \underline{\mathrm{then\ TRUE}}\\
&\qquad\qquad \underline{\mathrm{else}}\ (\forall \beta \sqsubseteq \alpha)(\underline{\mathrm{ORDLIST}}(\beta)[P]\ \&\\
&\qquad\qquad\quad [\underline{\mathrm{head}}(\beta),\ \underline{\mathrm{head}}(\alpha) =: x_4, y_4]\underline{\mathrm{HOD}}.\ \leq)]);\\
&\qquad\qquad\qquad \underline{\mathrm{def\ FALSE}}: 0 = 1;\\
&\qquad\qquad\qquad \underline{\mathrm{def\ TRUE}}: 0 = 0;\ \underline{\mathrm{end}}.
\end{aligned}$$

Extending the notion of Σ-definition (allowing Σ-schemes to be referred to in Σ-formulas), we thereby have extended also the notion of a Σ-scheme itself. This also enables us to introduce the concept of a Σ-program.

Let J be an (open) Σ-scheme and Q its formal n-place predicate parameter (the case of a functional parameter will be considered similarly). We consider a Σ-scheme $J1$ and its input $S_i(\overline{x}_i)$ with formal parameters $\overline{x} = \langle x_1, \dots, x_n \rangle$. An expression of the form $\langle [J1.S_i(\overline{x}) =: Q]J; J1 \rangle$, denoting the substitution of the expression $J1.S_i(\cdots)$ for each occurrence of Q in J, will be called a *J1-instantiation* of the Σ-scheme J for the formal parameter Q. We will say that the Σ-scheme J is *concrete* if all its formal predicate and functional symbols are instantiated.

DEFINITION 5. A concrete Σ-scheme J is called a *Σ-program* if all Σ-schemes involved in its instantiation (called the *Σ-subprograms* of J) are also concrete Σ-schemes.

It follows from the definition that, for example, every closed Σ-scheme is a Σ-program. And again: just as in the case of closed schemes, we can show that the following holds.

THEOREM 7. *Every Σ-program defines a monotone operator, whose least fixed point is Σ-definable.*

In speaking of J as a Σ-program, we thereby emphasize that not only the Σ-scheme J is under consideration, but the whole complex of concrete Σ-schemes involved in its definition. Some of these Σ-schemes may be *standard* (built in) constructs of the Σ-language. A concrete sequence of standard Σ-programs determines to a great extent the convenience and the problem orientation of the ΣS-language.

The initialization of a Σ-program J proceeds in the following way. Let $S_i(\overline{x}_i)$ be an input of J (which includes the inputs of any of its subroutines). It is assumed that all the names of Σ-programs involved in the definition of the Σ-program are unique. Let In and Out be reserved words denoting input and output data. We divide the list $\overline{x}_i$ into two disjoint sublists $\overline{y}$ and $\overline{z}$.

Then the expression

$$\langle \underline{\text{In}}\colon \overline{a} =\colon \overline{y};\ \underline{\text{Out}}\colon \overline{z}; J.S_i \rangle$$

or

$$\langle \underline{\text{In}}\colon \overline{a} =\colon \overline{y};\ \underline{\text{Out}}\colon \overline{z};\ J.J'.S_i \rangle,$$

where J' is the Σ-subroutine in which the input S_i lies, is called an *access* to the Σ-program J.

We say that the Σ-program J *computes* the value of a variable of z in the model $\mathfrak{A}(\mathfrak{M})$ under the access $\langle \underline{\text{In}}\colon \overline{a} = \overline{y};\ \underline{\text{Out}}\colon \overline{z};\ J.S_i \rangle$ if there is a sequence of elements $b \in |\mathfrak{A}(\mathfrak{M})|$ such that

$$\mathfrak{A}(\mathfrak{M}) \models \Phi_{\Gamma_{\Phi_i}}(\overline{a}, \overline{b}),$$

where Φ_i is the corresponding Σ-formula in the definition of S_i, Γ_{Φ_i} is the monotone operator corresponding to this formula, and $\Phi_{\Gamma_{\Phi_i}}$ is the definition of its least fixed point. The question as to how the computation of the Σ-program proceeds will be the subject of later publications.

To conclude this article, we would like to mention that the authors did not set themselves the goal of an exhaustive investigation of the concepts of Σ-programming. The goal was simpler: to demonstrate the theoretical possibility of constructing a language and a corresponding programming style on new, logical foundations, different from the traditional, conventional ones. There remains much work to be done in the further development of the ideas of Σ-programming, to convert them into a practical working system.

We wish to direct the reader's attention once more to the fact that all the constructions introduced above (Σ-definitions, Σ-schemes, Σ-programs) have a list structure and therefore may be considered themselves as data for the Σ-language.

Thus the Σ-language is not only a language for the construction of Σ-programs, but also a language in which these programs can be transformed, their properties investigated, formal constructions described which arise in the design of Σ-programs, etc. The Σ-language is a high-level language oriented toward work with complex information with a logical structure.

References

1. Jon Barwise, *Admissible sets and structures*, Springer-Verlag, 1975.

2. Yu. L. Ershov, *The principle of Σ-enumeration*, Dokl. Akad. Nauk SSSR **270** (1983), 786–788; English transl. in Soviet Math. Dokl. **27** (1983).

3. ——, *Dynamic logic over admissible sets*, Dokl. Akad. Nauk SSSR **273** (1983), 1045–1048; English transl. in Soviet Math. Dokl. **28** (1983).

Translated by G. L. CHERLIN